$8.95

R2R

PANIC IN
THE PANTRY

PANIC IN THE PANTRY

Food Facts, Fads and Fallacies

BY

Elizabeth M. Whelan, Sc.D., M.P.H.

AND

Fredrick J. Stare, M.D.

New York ATHENEUM 1975

*For Those of Us
Who Look Forward
to Meals*

January 3, 1974

Food and Drug Administration
Washington, D.C.

Dear Sir:

You are always taking something off the market because it is harmful to us.

Well, we all know that we have to have salt. But I hear that in excess amounts it is also harmful.

Are you going to take salt off the market too?

Please advise.

> *Sincerely yours,*
>
> *Marion L. Cooper*
> *Akron, Ohio*

March 5, 1975

Food and Drug Administration
Washington, D.C.

Gentlemen:
I have recently read a story about a purple coloring which might cause cancer.

If it does, I would appreciate your letting me know about it as soon as possible.

> *Very truly yours,*
>
> *John Anderson*
> *Yonkers, New York*

(ACTUAL LETTERS TO FOOD AND DRUG ADMINISTRATION, NAMES CHANGED TO PROTECT THE PANICKED.)

Preface

IT ALL began with the banning of cyclamates in the fall of 1969. That's when I first began to wonder if eating was hazardous to my health.

For ten years I had been enjoying the noncaloric sweet life. I shook a few drops of a miracle liquid into my coffee each morning, quenched my thirst at lunch with a chilled bottle of calorie-free soda, and garnished my dinner salad with a thick, creamy, and non-guilt-provoking diet dressing. Over the years, I read newspaper and magazine stories about cyclamates and saccharin and I was convinced of their safety. I had nothing to worry about. Except watching my weight.

Then, almost out of nowhere, the Secretary of Health, Education and Welfare informed me (and millions of other cyclamate users) that the innocent-looking white powder that was perking up my low-cholesterol, low-fat breakfast brought about a cancerous tumor in the bladder of some unfortunate sweet-toothed rodents. Cyclamates disappeared. And so did my feeling of confidence that the Food and Drug Administration was protecting me from life-threatening foods.

If cyclamates could be suddenly shown to be cancer-causing agents, what about other types of chemicals which are routinely added to food? I started to read packaging labels compulsively. Butylated Hydroxyanisole (BHA). Butylated Hydroxytoluene (BHT). Sodium Bisulfite. Lecithin. Xanthan Gum. How did I know they were safe? The fact that I couldn't even pronounce them didn't make me feel any better.

Eventually I switched from reading labels to reading books. Authors informed me that there were poisons in my food—mischievous chemicals lurking in my cupboards, waiting for the opportunity simultaneously to pollute my "inner environment" and to scramble the genes of the next generation. A very unappetizing evaluation indeed. And what did these writers suggest? A return to nature, the consumption of foods that were "100% pure, organically grown," packaged with "no artificial anything." Natural foods, they said, were my key to a long and happy life and, most important, a means of lowering my odds on contracting the most dread of all diseases, cancer.

I was concerned. So I gave it a try. For three months I shopped at the Mother Nature Spa around the corner. I bought the makings for cold lentil salad, organic omelettes, fava bean casserole, desiccated liver stew, and cucumber-yogurt soup with dill. I drank so much Tiger's Milk that I thought I detected the shadow of a stripe pattern on my skin. I ate so much honey and granola that I began to feel like a sticky wheat product. I tried the natural way. I did not like it. And it was expensive. But I was still worried about food additives.

This sincere concern over the possible ill effects of food additives led me to delve deeply into both popular and scientifically oriented material on the "natural" vs. "artificial" controversy. I took a close look at *why* consumers were panicking in their pantries, *why* they were becoming increasingly suspicious about additives, and what specific factors were underlying their concerns. Then I examined food fads of the past to see if our current back-to-nature binge had any historical precedent. I talked with Food and Drug Administration officers, political representatives, independent scientific researchers, and members of the various industries that produce food chemicals. I questioned them about the validity of the flight from food additives, and about the current safety of products at our local supermarket. I gathered some opinions about the laws that dictate our food regulation policies, and I learned how the food laws in general, and the so-called Delaney anticancer clause in particular, were applied in recent decisions to ban additives.

As a result of my personal inquiry into the food additive question and the back-to-nature movement, I reached four conclusions:

First, the current concern that the food additives now in use pose a health hazard is totally unfounded. Second, the flight from food additives to purely "natural" products may lead to some very serious health problems, to say nothing of the extra drain on the pocketbook. Third, while food fads and concerns about food safety are nothing new, they are today receiving an extra boost from the Delaney clause, a controversial piece of legislation that promotes unscientific and panic-based bannings of food additives. Fourth, the contemporary back-to-nature mania, like other similar movements throughout history, is a hoax, perpetuated by opportunists who are intent on taking advantage of nervous and very gullible consumers. And having arrived at my own personal verdict about additives, I could return to enjoying "regular food."

But I saw people all around me who were still panicking in their pantries about "artificial" chemicals. So I turned to one of our most prominent contemporary nutritionists, Dr. Fredrick J. Stare, Professor of Nutrition and Chairman of the Department of Nutrition at the Harvard School of Public Health, and a former teacher of mine, and asked him if he would be interested in writing a book with me, a book for those consumers who are perplexed and concerned, one that answered some of the questions now being raised about the safety of our food. He agreed—and that's what *Panic in the Pantry* is all about.

What concerns us most about the current panic over food additives is that the basic facts about nutrition and health which have taken hundreds of years to accumulate are being overlooked in favor of isolated, scientifically irresponsible—but well-publicized—claims about alleged "poisons" in our food. Very often it is the same individuals who keep raising questions about food safety and, despite an abundance of scientific material to the contrary, lead American eaters to lose confidence in the ability of the Food and Drug Administration to keep our food safe. Obviously we should be interested in the quality of our food supply and we must have an effective, ongoing monitoring system to ensure that what we are eating is, by the most recent standards, safe and health-promoting. But this process should be a rational one, not one that makes our stomachs churn each morning as we read the headlines about "new evidence of hazards." And it should be a regulatory process that is free from political influences and the priorities of any particular interest groups, be they industry or consumer activists.

We knew from the beginning that the food additive question and the contemporary fascination with so-called natural foods was a highly controversial and emotional one. Concerns about food safety have always been on this level. We also knew that the current debate about additives has centered on individuals with two allegedly incompatible outlooks: those who are personally involved in the food industry and thus committed to defending additives and food-processing techniques and those on the "other side" who are suspicious of any industrially produced commodity, yearn for a return to "chemical-free living," and possibly are eager to promote their careers or the sale of their own "100% natural" products.

In the past, most material on the food additive question has been written by people who were professionally and personally aligned with one or the other of these two points of view. But in writing *Panic in the Pantry* we had neither of these "causes" in mind. We are speaking without malice to any individual or society. Instead, we are presenting the facts as we see them after having thoroughly and scientifically examined the subject. We were interested only in the facts about what is and what is not acceptable for the human diet and what is and will be the safest and most effective way of feeding the world's growing population. We reached our own independent conclusions with regard to the safety of food additives currently in use. And we would like to share these views with you.

I hope you find the chapters that follow both palatable and easy to digest. *Bon appetit!*

> Elizabeth M. Whelan, Sc.D., M.P.H.
> New York
> March 1975

Acknowledgments

WE'D LIKE to offer special thanks to a few of those individuals who were particularly helpful in gathering and evaluating the information that makes up this book.

Specifically we are indebted to Dr. Jean Mayer, Professor of Nutrition at the Harvard School of Public Health; Dr. Richard L. Hall, Vice President (Research and Development) of McCormick and Company, Inc.; Peter Barton Hutt, General Counsel of the Food and Drug Administration (FDA); Donald A. Berreth and Mary Carol Kelly of the FDA Press Office; Mary Hoban of Nabisco, Inc.; Peggy Walton, Manager of Consumer Information for the Manufacturing Chemists Association (MCA); Dr. Virgil Wodicka, formerly Director of the FDA's Bureau of Foods; Dr. Elizabeth K. Weisburger of the National Cancer Institute; Jelia Witschi of the Department of Nutrition of the Harvard School of Public Health; Howard W. Mattson, Director of Public Information at the Institute of Food Technologists; George W. Ingle and Morgan M. Hoover of MCA's Food, Drug, and Cosmetic Chemicals Committee; George Rinehart of *Nutrition Today*; U.S. Representative James J. Delaney of New York; Dr. Judith Goldberg of the Health

ACKNOWLEDGMENTS

Insurance Plan of New York; Ann Partlow; Martha Chapple Steel; Stephen T. Whelan; Joseph F. Murphy, Executive Vice President, The Continental Corporation; Marion L. Murphy; Marilyn Bartle; Sidney Harris; Paul J. McGeady; and Helen McGoldrick.

We also would like to thank the staff of the New York Academy of Sciences Library for their efficient and courteous services in locating the vast amount of research material that was vital to the preparation of this book.

Introduction

A LONG overdue book has finally been written about food additives, by authors who are professional nutritionists. They have achieved a remarkable result: a book that is understandable, interesting, and truthful. One author, Dr. Elizabeth M. Whelan, is well trained in epidemiology and statistics and, therefore, is particularly qualified to interpret scientific data related to public health. She is a former student of the other author, Dr. Fredrick J. Stare, who as a physician and scientist has spent his professional lifetime in the field of nutrition. He is particularly qualified to write and comment about food additives and human health. As a team, these authors address the subject of food additives in great style. The book is refreshing, too, in that it exposes myths that are unsound nutritional practices and, therefore, beliefs that can cause problems in human health.

This book should be required reading in nutrition courses at both the college and high school level and, because it is well-written, should be read by laymen and health professionals alike, including physicans. A noteworthy achievement in the book, which I like, is its humor. We should not lose our sense of humor even when dealing with serious problems that affect our daily lives.

Considering that the toxicants in foods that are life-threatening are the naturally occurring toxicants, one wonders why so much fuss about food additives. This is explained in the book and, since the fuss affects our eating habits, we should learn the truth about additives. This book puts problems about food additives into proper perspective.

The book traces the history of the development of food additives and the origins of different food cults. Defects in these philosophies are expertly pointed out. As noted in the text, there are problems with food additives (an opinion I share), but the problem is under good control in our country, thanks to the FDA and academic laboratories.

What is the Delaney clause? The authors feel that it should be stricken from the law. If you do not know what the clause is or its impact, you will, after reading this book and, most likely, will agree with the authors that it should be so stricken. Some food additives are unique; certain foods must have them; no replacements are presently known. It is ironic that many people who want such additives taken out of food for no good reason are the very same people who ascribe health and increased life span to huge doses of vitamin, mineral, and other "super" preparations. I feel, however, that longevity will not be increased by adding these concoctions to our diet, but rather by moderating our caloric, protein, and other nutrient consumption and increasing our exercise. We know that both fat- and water-soluble vitamins can be toxic. The saving grace for many in our society who take these potent substances is that their bodies excrete them before damage to body organs can occur (not true for everybody, but for most, thankfully). The dangers of these preparations are well presented in this book.

Drs. Whelan and Stare also discuss food coloring and food preferences. They note, too, that industry and academic laboratories are especially sympathetic to the needs and wants of the public, if the desires are not harmful. Some changes, although off in the future, are coming and they will be possible because through food science, new products that are safe for human consumption can be developed.

Enjoy this book. It is one that deals with food additives in the best manner I've seen and also develops a sensible discussion concerning future feeding problems not only in this country, but around the world. The book is required reading for anyone interested in nutrition—and who isn't?

Ralph A. Nelson, M.D., Ph.D.
Head, Section of Clinical Nutrition
Mayo Medical School
Mayo Clinic

Contents

PANIC IN
THE PANTRY

ONE

From "What's for Dinner" to "Name Your Poison"

> *"The king ate enormously, stuffing the meat into his little mouth with his knife. As he munched, the meat and vegetables popping from cheek to cheek, his eyes shone with happiness."*
> A COMMENTARY ON HENRY VIII

BACK WHEN EATING WAS FUN

AN enjoyable, relaxing meal is one of life's greatest pleasures.

Certainly part of that pleasure is the gratification expressed by a previously empty belly. But for those of us who look forward to meals, eating is far more than a means of satisfying a biological need. It is, in addition, an intensely social, psychological, and—quite literally—a sensational experience. Not only that, but it's fun! Or at least it used to be.

Quite obviously we have to eat to keep our physiological machinery in operating condition. The drive that

3

ensures that we do eat is hunger, often described as "the great motivator of the human race." When an individual is really hungry he doesn't ask the source of the food he is being offered and he does not worry about its purity or quality. Hunger is a drive that will overrule any rational consideration. History is full of reports of how starving people ate dirt or fought over a limited supply of raw shark meat in a desperate attempt to survive. During the early famines in England, the food supply was so threatened that a good rat, dead or alive, was worth a shilling. Hungry people made the observation that

> Rats are not a dainty dish to set
> before a king,
> But for a really hungry man,
> they are just the very thing.

The "I'd rather starve first" attitude has little historical support. Hungry people eat what is available—or, as the saying goes, "the famished belly has no eyes."

HUNGER VS. APPETITE

Most of us are unfamiliar with the real hunger drive. Living as we are, surrounded by foods of many types, we can pick and choose exactly what we want to eat. And because we eat relatively frequently, the force that determines what we choose is more likely an appetite pang than a hunger pang. For us, it is the nonnutritional extras that separate the physiologic drive of hunger from the mental and emotional associations that characterize the word *appetite*.

The human stomach has been described as "the world's most delicate test tube." In a matter of seconds, a previously healthy appetite can disappear. Consider, for instance, a situation where you are invited to a friend's home for dinner. You arrive and announce that you could eat a horse. The question however is, Can you eat the appetizer set before you (fried caterpillars) and the gourmet Chinese-style main course (tender young white mice dipped in honey)? Chances are you can't. Almost instantly that enormous "hunger" you brought with you disappears. You have "lost your appetite"—at least temporarily.

We are social beings, born into cultures that determine our response to specific foods and the way they are served. For us, the appearance of food, its palatability, and the surroundings in which it is served must be right before we can enjoy a meal. A fly on the butter, a hair in the gravy or, as was the case above, a culturally unacceptable menu is more than enough to turn what could have been an enjoyable experience into a fierce battle between your gastric juices and the antacid you just took.

There's another factor that can have a similar unsettling effect: a fear that your food is unsafe. How could you possibly enjoy a meal if, with every bite, you thought you were slowly poisoning yourself to death?

CHEMICALS!

Consider a typical consumer who enjoys a soup course before each dinner meal. To save time and energy, she chooses one of the instant dry soups on the market today. It's simple to prepare, there's no pot to clean, and everyone enjoys it. One night during the dinner conversation, a family member points out that he has been hearing on the radio that many of the "chemicals in food" are dangerous. In the middle of the soup course, he demands to know exactly what is in this creamy delight in front of him. A look into the pantry reveals that the cream-style chicken-flavored soup contains:

Spray-dried vegetable fat (vegetable fat, corn syrup solids, sodium caseinate, mono and diglycerides, dipotassium phosphate, sodium silico aluminate, artificial flavor and color), food starch modified, whey solids, salt, dehydrated chicken, monosodium glutamate, nonfat milk solids, vegetable gum, buttermilk solids, malto-dextrin, hydrolyzed milk and vegetable protein, hydrogenated vegetable oil, dehydrated onions, dehydrated parsley, wheat starch, yeast extract, chicken fat, potato starch, corn starch, flavorings, and turmeric.

A veritable army of nasty chemicals! All of a sudden that liquid in the soup bowl becomes considerably less appetizing.

It's not just instant soup that people are losing their appetites about today. A significant portion of Americans is sincerely concerned about the safety of the food they eat—and worried about all those "chemicals" in our diet. This concern is making eating a less enjoyable experience. Where once they could shop in a carefree manner, efficiently dropping various foodstuffs into their carts as they moved from aisle to aisle, many shoppers are now moving very slowly through the market, cross-examining package labels as they go along. And many of these consumers are both suspicious and concerned enough to write to the Food and Drug Administration either to express their indignation about what they perceive to be a lack of consumer protection or simply to ask questions about the safety of particular

"I'm so glad you like it. Actually it's just sodium acid pyrophosphate, erythorbate, and glucono delta lactone with some meat flavoring."

food additives. Each year, by means of telephone, telegram, and letter, over 200,000 American consumers do just that. A sample of one day's mail follows.

"DEAR FDA, I'M CONCERNED..."

Dear Sir:

Enclosed please find some crackers. I am concerned about what is in them.

Would you please examine them and report directly to me?

Sincerely yours,

Gentlemen:

Why are the following put into foods: a) calcium chloride, b) calcium silicate?

What harm can they do to the body? I feel silly asking you this since I really know that if the FDA allows it, it can't be bad.

Very truly yours,

Dear Mr. FDA:

It takes you about twenty years to find out if a food causes cancer. It takes about half that time to have cancer wreak havoc in our bodies. I'd rather starve.

Very truly yours,

To whom it may concern:

With all the unnecessary food additives today, no wonder everyone is ailing in some way or other. When we were kids, they used to have the lettering "100% pure" on all foods. Why don't they now? I know those additives aren't doing us any good.

Yours,

P.S.   Would you please check out apple jelly (Finast brand)?

Gentlemen:

I have written to Atlanta and Ralph Nader about meat tenderizers.

I have been poisoned five times and have had to go to hospitals. My doctor says it is the meat tenderizer that is affecting my intestines.

Will you please look into this matter?

Sincerely,

Dear Sir:

I am enclosing a package of vanilla cookies. Is ammonia allowed in human consumption?

Sincerely,

———————————

Dear Sir:

I am concerned about the warning label that now appears on Tab and other diet soft drinks.

When I am pregnant (hopefully next month) should I drink Tab?

Very truly yours,

———————————

Dear Sir:

I am in school, and I just heard about what saccharin can do to you if you eat too much.

I was chewing some bubble gum and my teacher asked to see the wrapper. She saw it had saccharin and she told me that it would cause cancer.

I wish, therefore, you would take everything that contains saccharin off the market.

Thank you.

Sincerely yours,

———————————

Dear Sir:

I like cod, kipper, salmon or anything, but I don't like the coloring matter they put in. Can you order them to leave out the color?

Yours,

———————————

9

Dear Sir:

Enclosed please find four pieces of Aunt Martha's delicious candy.

I am eighty-six years old and must leave sugar out of my diet, and I feel that the label on this package is misleading.

Thank you,

P.S.   I am a retired chemistry teacher.

Dear Sir:

I have been feeling very depressed recently, and I suspect that is because of all the artificial additives I have consumed in the course of my long lifetime (seventy-one years).

Would you please send me references you might have linking chemical additives and mental disorders.

Sincerely,

Dear Sir:

You are always taking something off the market because it is harmful to us.

Well, we all know that we have to have salt. But I've heard that in excess amounts it is also harmful.

Are you going to take salt off the market too?

Please advise.

Sincerely yours,

Gentlemen:

I have recently read a story about a purple coloring which might cause cancer.

If it does, I would appreciate your letting me know about it as soon as possible.

Very truly yours,

Gentlemen:

The enclosures are to call your attention to chemicals which are used to make bread white, softer, and to keep longer.

Any concerned housewife is not interested in whiteness or softness of the bread she eats or serves to her loved ones. She wants good bread—a natural product, one which would be good for human beings.

Sincerely yours,

Dear FDA:

Enclosed is the top of a container that was supposed to contain coffee cream. Can you tell me what the liquid really is? I cannot read the chemicals without a magnifying glass.

Regards,

Dear Sir:

Would you please tell me if the enclosed raisins contain any preservatives to retard spoilage?

Very truly yours,

Enclosed: raisins

Dear Mr. Secretary:

As I sit here pondering the ingredients of this bottle of blue cheese dressing, I am again faced with a baffling question. Although the company's good enough to print what is in the dressing, I am still confused.

What are potassium sorbate and monosodium phosphate?

Yours truly,

P.S. I am suspicious.

Dear Sir:

I have recently sought to substitute an artificial sweetener for sugar, but I have been dismayed to find that all sweeteners carry a warning "should be taken only by those on sugar-restricted diets."

Now I am not on any diet. But I read awful things about sugar and how dangerous it is.

What I would like to know is what can happen from eating sugar or artificial sweeteners, and which is the lesser of the two evils?

Sincerely yours,

P.S. I hope the FDA has some idea of the answer.

To Whom It May Concern:

I am writing to ask you if you can help us with a perplexing problem we are having with our daughter.

I have been reading of the dangers of saccharin. My daughter does consume a few bottles of soft drinks a day and has begun to pass stones. I wonder if it is the result of the saccharin. Could Vicky's stones be caused by saccharin?

Very sincerely,

Gentlemen:

Just read that DES is back on the market. I have a news release that goes on and on about how the FDA was finally convinced it was bad. Now you reverse.

After again thanking you for protecting our health (ha!) I only have one last thing to ask and from now on I shall try to repeat myself only once and not get my blood pressure up about what is going on: Please start labeling the meats and fowls so we can choose ones without DES!

PANIC: IT'S A CHEMICAL REACTION

These people—and many of the rest of us—are obviously worried! Indeed, many of us are panicked and very angry at the FDA for what we feel is the agency's inexcusable failure to keep our food safe. And each day our concern and anger intensifies. Radio advertisements tell us that "common food additives" may be responsible for depression and other forms of mental duress.

Magazines such as the Rodale publications, *Prevention*, and *Organic Gardening and Farming* and even an occasional physician contend that the chemical additives in food are responsible for hyperkinesis (a behavorial disorder characterized by excessive physical activity and an inability to concentrate and learn), and that the "FDA wants you to hand in your brain." A long list of unappetizing books—*Our Daily Poison* (Leonard Wickenden, 1955), *The Poisons in Your Food* (William Longgood, 1960), *Two Hundred Thousand Guinea Pigs* (Fuller, 1965), *The Hidden Assassins* (Mooney, 1965), *How to Live in a Poisoned World* (Bethel, 1969), *The Chemical Feast* (Turner, 1970), *The Deadly Feast of Life* (Carr, 1971), *Food Pollution: The Violation of Our Inner Ecology* (Marine and Van Allen, 1972), *Body Pollution* (Hull, 1973), *Eating May Be Hazardous to Your Health* (Verrett and Carper, 1974), *Why Your Child Is Hyperactive* (Feingold, 1975)—claim that mad scientists in Washington are poisoning our food and that there is a cancer-causing agent on every dinner plate.

Newspapers recently carried stories about "Food Day" and the so-called "terrible foods" in the American diet. Specifically, Dr. Michael Jacobson, Co-Director of the Center for Science in the Public Interest in Washington, D.C. identified the top Terrible Ten (which "exemplify what's wrong with the American food supply") as table grapes (the focus of a boycott by the United Farm Workers), Coca-Cola (because "it contains no

BOOTH

"There is no nourishment in this meal, but then, by the same token, there is a minimum of poisons."

© 1974 by The New Yorker Magazine, Inc.

nutrients and costs more than milk"), Breakfast Squares (termed "Gainesburger for people"), bacon (because it has the additive nitrite), Wonder Bread (because it is manufactured by a division of ITT and ITT owns Sheraton Hotels and makes military supplies), sugar (because it allegedly is the cause of certain health problems and sugar refineries have been charged with price fixing), Gerber Baby Food Desserts (termed "junk food"), Frute Brute (charged with having unnecessarily high price tags and sugar content), Prime Grade Beef (because it is "fattened in feed lots on grain that could otherwise be consumed by hungry people") and Pringles ("the ultimate insult to the potato").

It is easy to single out and accuse food categories— even if you don't have any facts to back you up. Perhaps it is also entertaining for those who are allegedly working in the "public interest" to sit around a table, draw up indictments for specific foods and then sit back to watch the turmoil that follows. But it is certainly not a useful approach to food regulation and can do nothing but inhibit our attempts to increase the availability of a variety of wholesome, nutritious, convenient, attractive and enjoyable foods. Indeed, the sweeping condemnation of those ten food categories is unrealistic and scientifically irresponsible given that there are no facts to support the claim that we would be better off if we excluded all of them from our diet. To the contrary, these foods, like *all* foods available today, when used in moderation and in the context of a well-balanced diet *contribute* to both our physical and psychological well being.

We are literally being swamped with scare stories about our food. Richard H. Barger, Executive Vice President of Agripac, Inc. has described the poisons-in-your-food campaign as "whipped up froth of ferment" that is "breeding concern about the safety and wholesomeness of our food supply." And in a presentation before the 1975 meetings of the American Association for the Advancement of Science in New York City, Dr. Bernard L. Oser pointed to many instances in which "the application of conjecture or hyperbole to fragmentary and inconclusive data" has resulted "in exaggerated concern for the safety of foods." He expressed the concern that "crying wolf too often will . . . detract attention from real dangers as has been happening in the use of tobacco and drugs, particularly by the young." But still the forms of faddism and so-called "nutrition education" go on, and many gullible eaters are taken in by it.

But why all this panic about various foods and food additives? Why are some people rushing to Healthfoodland to buy "safe" products, and why are others eating "regular" food but enjoying it less? And how did such a dramatic shift in opinion occur almost overnight?

It is clear that there are two important and related factors that have provided the ideal background for the flight from food additives to Healthfoodland. First, our widespread concern about cancer and some other dread diseases, and second, our current form of food legislation which encourages an "eat it today, ban it tomorrow" approach. Both these pressures behind our eating-natural craze are further reinforced by the misinformed (or maybe simply unscrupulous) people who perpetuate the common—yet completely inaccurate—assumption that only "artificial" foods contain "chemicals" and are suspect, while "natural" foods are "chemical-free" and by definition good.

CANCER

As you will read in the next chapter, people throughout history have been deterred from eating certain foods because of vague rumors that they are linked with a feared disease. And the situation is no different today. Cancer, for instance, in its many forms is now one of the most prevalent and least understood of all diseases. Its incidence has increased significantly since 1900 and it is now the second leading cause of death in this country (just behind heart disease).

Since we don't understand the origin of this disease, everything in our environment is suspect. But in the context of that "chemical vs. natural" dichotomy, additives and "unnatural" eating have become primary suspects in the cancer "who-done-it" mystery.

Imagine the situation. You're told cancer has moved from being the eighth leading cause of death in 1900 to the second leading cause in 1970. Then you're informed that the use of food additives has also increased in this same seventy-year span. Conclusion? Following the *post hoc, ergo propter hoc* reasoning, many have concluded that cancer is linked with use of "unnatural" food chemicals.

At first it may sound like a tenable hypothesis. But then, if you look a little bit more closely at the figures, you'll see it is not supported by facts. Statistics clearly indicate that *the rise in cancer deaths in the United States in the past forty years can be attributed to an increase in lung cancer:* the lung-cancer death rate is now eighteen times as high for men and six times as high for women as it was forty years ago. There is just no way to link food additives with lung cancer. The greater frequency of lung-cancer mortality is directly related to the growth in popularity of cigarette smoking between 1900 and 1964. As is evident in the chart below, the frequency of cancer from all other sites—for instance, the stomach, which one might suspect could be affected by food—has

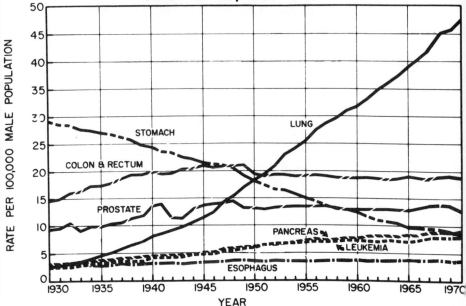

MALE CANCER DEATH RATES* BY SITE
United States, 1930–1970

*Rate for the male population standardized for age on the 1940 U.S. population.

Sources of Data: National Vital Statistics Division and
Bureau of the Census, United States.

declined or stabilized. Indeed, as will be discussed in a later chapter, food additives may be responsible for all or part of the decline in stomach-cancer mortality.

But the cancer scare and its alleged link with non-natural foods is enough at least to lay the groundwork for a shift away from "chemicals."

EAT IT TODAY, BAN IT TOMORROW

But again, the fear that foods may be linked with diseases has always been with us. The concern about cancer does not alone account for the current preoccupation with 100% natural products. Rather, this type of previously latent fear has been vividly intensified by the Food and Drug Administration's recent banning and "reevaluation" spree.

It all began with the precipitous recall of cranberries and mushrooms in the 1960's. But what really set off the panic was the unprecedented banning of the artificial sweetener cyclamate in the fall of 1969. In what has been described as a mood of "cyclamania," the Secretary of Health, Education and Welfare dramatically announced that the artificial sweetener which many of us had grown to love was no longer allowed in sugar-free soft drinks and the large assortment of diet foods we'd been eating for the past few years. Why? Because eight rats who had been stuffed for two years with unbelievable amounts of cyclamate, saccharin *and* in some cases another chemical had developed tumors in their bladders. "Cancer" was the lead word in most of the headlines. And before we could even say "sweet and low," all the food products and drugs containing cyclamates were whisked from our shelves. That was enough to make anyone's sweet tooth ache—and it was also enough to make us all wonder what other foods could be "linked with cancer" and whether all additives were "badditives."

Within a year or so after the cyclamate ban, a certain color additive and the cattle-growth stimulant DES were banished from our kingdom. And saccharin, the only sugar substitute left; MSG, the flavor enhancer; and nitrate and nitrite, the meat- and fish-curing agents, were put on trial. With all this sudden concern being expressed about the safety of products that had been used for some twenty years (as was the case with cyclamates) or even for hundreds of years (as was the case with MSG and nitrates), it is no wonder we became suspicious about the food we were eating. The prevailing questions were,

"If cyclamates, DES, saccharin, nitrates, nitrites, MSG, and some food colors are all of a sudden unacceptable or under suspicion, what about all these other strange-sounding chemicals that go into our food? How do we know they won't be banned tomorrow? What *is* safe for us to eat?"

It was with the banning of cyclamates in 1969 that the panic-in-the-pantry phenomenon took hold. And with each new banning or "reevaluation," that panic intensified. Underlying our concern with additives—and the associated natural-foods fascination—is the belief that the *restrictive actions taken by the Food and Drug Administration were based on rational scientific information.* In other words, we thought the FDA was acting independently, without pressures from specific interest groups. In the case of cyclamates, for instance, most of us assumed that the sweetener was banned because there was a significant amount of medical evidence indicating that it posed a health hazard to humans. Indeed, as will be pointed out later, the majority of consumers were actually *pleased* that the FDA was "taking good care of us"! What they, and most of us, did not know at the time was that these bannings and subsequent reevaluations were occurring not because of any legitimate medical and scientific data, but because of the existence of a certain very controversial and politically sensitive piece of legislation known as the "Delaney anticancer clause."

The Delaney clause, part of the 1958 food additives amendment to the Federal Food, Drug and Cosmetic Act of 1938, states that any additive which is found to induce cancer when ingested by man or animal must be removed from the market. There are no ifs, ands, or buts about the clause. If an additive, when ingested by any animal, in any dose, for any length of time, brings about any form of cancer in even one experiment, the product must be banned. The Delaney clause does not apply to a "natural" product, even if Mother Nature's own delight brings about the same type of cancerous growths at significantly lower dose levels. The immunity of these "natural" products from the wrath of the anticancer clause serves to reinforce the popular yet unfounded belief that only "artificial" substances could have nasty "chemicals."

You'll read more about the Delaney clause and how it has been used to condemn or call into question some innocent—and very useful—additives. But what is important here is the identification of the clause as a major force behind the flight from additives. Without this piece of legislation, we would still have food fads—but they would not, as is the case now, involve major portions of our population.

WHAT'S THE EFFECT?

The most obvious effect of the flight from food additives is the extraordinary proliferation of the so-called health food stores, those that claim to offer products that are not only "100% natural," free from any artificial anything, not only produced using "organic" methods but also filled with vitamins, minerals, and other supplements which have been described as having an almost magical potential. These products are promoted first by attempts to further discredit "normal" food and second by advertisements that link "health" products with increased chances for a long and disease-free life.

So what's wrong with that? How could a little food fad hurt us? There's always a fringe element in any society, so why don't we just ignore it and hope it goes away?

But there are some potentially very serious problems.

First, there is the concern about the impact exclusively "natural" living can have on health. It is obvious that any type of exaggerated food faddism can cause serious health problems or even death. The recent example of a twenty-four-year-old girl who died after losing fifty pounds as a result of following a "healthy" macrobiotic diet demonstrates this point. But the trend toward additive-free food may have more far-reaching effects if, as is discussed in a later chapter, major food producers respond to public pressure and omit preservatives and food fortification agents.

Second, and related, an increase in popular concern about food additives and the fascination with natural-organic products has the potential for presenting a threat to the maintenance and expansion of our already imperiled food supply. "Naturalists," for instance, condemn the use of pesticides. But it is a reality of modern life that some type of pesticide is necessary for efficient farming. As pointed out by Dr. Norman E. Borlaug, a Nobel Prize winner for his work in developing new strains of wheat, "If the use of pesticides in the United States were to be completely banned, crop losses would probably soar to 50 percent and food prices would increase fourfold to fivefold."

Similarly, Dr. Earl Butz, secretary of agriculture, estimated that 50 million people in the United States alone would face starvation if we were to depend solely on organic farming.

And as a decision to omit pesticides can reduce the crop yield from an acre of land, so can the omission of preservatives further reduce the amount of food available to consumers who cannot produce their own and must rely on market-bought food.

It is clear that eating natural is an economic luxury.

Gloria Swanson, who attributes her flow of energy to pure, organic foods, can afford to say, "I won't eat anything with artificial coloring. No dyes. No preservatives. No added chemicals. Just God-given food—that's all." And the dancer Allegra Kent can choose to eat only "eggs . . . the fertile kind from health food stores." But natural and organic living is not a viable alternative to worldwide food production, and indeed a continued fascination with additive-free food may undermine the advances in food technology that have accumulated over the past two centuries.

Perhaps, as Professor Thomas H. Jukes of the University of California at Berkeley has pointed out, modern health faddists cannot understand how precious food is, having themselves never had to risk drought, blight, and insects for the privilege of getting something to eat.

Third, and not as immediately obvious, a concern about food safety is threatening an age-old source of social, psychological, and physical pleasure: *Eating used to be fun.* But now we are all concerned. And what are our options? We can go to Healthfoodland, choose from a limited supply of very expensive foods—and try not to think about all those delicious goodies we used to eat. Or we can stick with our regular diet, and just worry— and, in the process of fretting about food, we can forgo the psychological pleasure that should be part of the meal. (Who could relax with a chilled Manhattan cocktail—or a bubbling diet soft drink—while glaring into the glass wondering if the maraschino cherry or the artificial sweetener was going to make him a cancer victim?)

But there is one other thing we could do. We could get the facts about this natural food craze and the panic about additives. Specifically, we might get the answer to five basic questions:

FIRST: *Where does our current additive-phobia fit in terms of historical perspective?* Is this back-to-nature binge really all that new? Are there other examples of food scares?

SECOND: *What exactly is the current natural-foods hoax all about?* Who is "eating natural," what types of products are they buying, and what are some of the problems associated with the current fascination with Healthfoodland?

THIRD: *Is there any validity to the basic premises of Healthfoodland?* If something is natural, does it by definition mean it is free from "nasty chemicals" and full of health-promoting materials?

FOURTH: *What are food additives?* What do they do and, most important, are they safe? Why were some very widely used additives recently banned? What would happen if we eliminated all additives from our foods?

FIFTH: *How are food additives regulated today?* What are some of the problems associated with food laws in general—and the Delaney anticancer clause in particular? How can these laws be changed to allow a more rational regulation of our diet—and to promote a return of our confidence about food safety?

The pages that follow attempt to provide some answers to these questions. Let's start by looking backward in time to consider some of the historical and hysterical food fads of the past and how they managed to capture the minds and stomachs of unsuspecting eaters.

TWO

Eating Naturally through the Ages

"There iz lots ov people in this world who spend so much time watching their health that they ain't got no time to enjoy it."

Josh Billings (Henry Wheeler Shaw) in Civil War Humor

MILK CAN KILL YOU!
(AND OTHER SAGE HISTORICAL ADVICE)

TODAY some people are concerned about the safety of "unnatural" food additives and the effect they might have on human health. But this concern is nothing new. Men have complained about the food served them since the time of Adam. In past years, food panics have focused on the safety of milk, meat, fish, coffee, vegetables, fruit, and a number of other items that are normal entries on our weekly shopping lists.

In the late 1940's, Dr. Melvin Page, a dentist, wrote *Degeneration-Regeneration* (published by himself in

25

1949), in which he strongly urged people to give up milk. According to Dr. Page, milk drinking was "unnatural," as man "and a certain species of ant are the only ones who use an animal secretion after the age of weaning." Milk was described as the underlying cause of colds, sinusitis, colitis, and cancer. As statistical evidence for his hypothesis, the dentist pointed out that Wisconsin, the leading milk-producing state, had a high rate of cancer deaths.

As a result of Dr. Page's admonitions, many followers stopped drinking milk and discontinued serving it to their children. They were frightened. They didn't want to eat or drink anything that carried with it even the remotest chance of being harmful to health.

Other foods have been likewise condemned for a variety of different reasons. A primary criticism of meat (apart from any underlying vegetarian or religious consideration) revolved around its alleged association with "sexual excess." In the South Pacific islands, a type of shellfish was proscribed during pregnancy to ensure that the child would be born without scales on its head. At one point, coffee went on the proscribed list when a rumor to the effect that it was colored with horse's blood convinced people that it was less than a heavenly beverage.

Potatoes and fresh fruits have long been suspect. In the British colonies, a popular superstition stated that if a man ate a potato each day, he wouldn't live out seven years. The belief that the ingestion of fresh fruit was linked with severe sickness, and even death, remained strong until into the nineteenth century. During the cholera epidemic of 1849, the *Chicago Daily Journal* recommended restrictions on all orchard produce. During the same year, the *Chicago Democrat* carried a story about two boys who, after eating some fresh fruit, went to the circus, only to meet with tragedy: "In a short time, one was a corpse and the other reduced to the last stage of cholera."

Still another report, which fanned the fire of the fruit scare, stated that a man merely passing a fruit stand filled with spoiled peaches suffered a severe attack of "the gripes."

People were panicked. They refused to eat fruit. Even after it was established that the link between peaches and disease was the merchants' habit of washing their products in polluted streams, there was, for many years, a vague but prominent belief that fruit was dangerous.

In addition to avoiding some allegedly dangerous foods, gullible listeners have been advised to change their eating styles as well: in 1920 some health lecturers warned against eating carbohydrates and protein at the same meal. Others pointed to the importance of avoiding meals that mixed acid and alkaline substances lest these antagonists frustrate one another's nutritional efforts. (No acknowledgment was made that all foods meet an acid environment as soon as they enter the stomach.)

A group of early twentieth century enthusiasts, headed by Horace Fletcher, warned that "nature will castigate those who don't masticate." Mr. Fletcher recommended that foods be chewed at least thirty-two times to ensure the absorption of their full nutritional value. He further opined that it was bad to swallow directly any food in a liquid state; it was first necessary to roll it around in your mouth a few times. Presumably there was a hidden benefit since, with all that chewing and swirling, one would eat less.

If you looked hard enough, you could probably find

historical expressions of concern and outright condemnation of most every item in the human diet. Foods are highly susceptible to rumor, and rumors promote faddism.

These lofty and pseudoauthoritative proscriptions of particular food items have two factors in common. First, they have no scientific basis. Second, they cater to the preexisting fears of the audience—and thus they don't require any scientific basis. The prospects of disease and death always have had enough fear-provoking connotations so that any "solution" will be considered by some portion of a population. If susceptible people hear an authoritative person state that milk causes cancer, they may say to themselves, "Maybe it does, maybe it doesn't, but why should I take the chance until we find out for sure?"

Fear interferes with an attempt to look at food-scare statements in a rational manner.

GARLIC WILL "CLEAN YOU OUT"!

The corollary to the belief that certain food and eating styles are bad for you is that specific others both prevent and cure disease—and generally promote good health. Garlic has long maintained its position as a leading superfood.

The ancient Dioscorides claimed that garlic was an excellent food both to clean out arteries and to open obstructions, although he warned that if taken in excess, it would stir up bodily lusts and become a source of lechery. The Egyptians fed large doses of it to their laborers to keep them healthy and invigorated as they were building the pyramids. And Adolphus Hohensee, during the 1940's, urged his audiences to put therapeutic faith in garlic as a means of curing low blood pressure, inhibiting germs, and cleansing their blood and intestines. Hohensee followers were urged to put a piece of garlic in their rectum at night. Allegedly they could expect to taste the flavor of this miracle worker in their mouths the next morning, and could then be assured that the full cleansing job had been accomplished while they slept.

But garlic wasn't the only popular "natural" purgative. Sarsaparilla, molasses, and sulphur (which "cheered while it cured"), as well as good old-fashioned seawater, were often recommended in April or May as a part of the general spring housecleaning. Very often these materials were used in conjunction with a period of fasting.

The idea that all food is generally poisonous, albeit a necessary poison, has been around for a long time. Historically, there have been those who felt that diseases such as smallpox and measles were merely manifestations of food-related "corruption escaping from the body." The Egyptians fasted for a full three days each month to rid themselves of these poisons. And in this century, at least two popular books have advocated periodic starvation as a key to health.

In 1908, Hereward Carrington wrote a 648-page book entitled *Vitality, Fasting and Nutrition*. A few years after that, Upton Sinclair, who was a chewing Fletcherite, followed up the Carrington work with his own book, *The Fasting Cure* (1911), wherein he recommended prolonged starvation as a means of combating tuberculosis, syphilis, asthma, liver trouble, and cancer. In a subsequent book (*The Book of Life*), Sinclair wrote: "I have known of two or three cases of people dying while they were fasting, but I feel quite certain that the fast did not cause their death, they would have died anyhow."

But staying healthy often involved more than garlic and periodic fasting. History books have identified a whole array of foods which allegedly offered special preventive and/or curative powers. Indeed, it is often difficult to distinguish between dietetic and pharmacological substances.

Pliny recommended cucumbers for "hot stomachs and hot livers," Egyptians fed sick children skinned mice, and Marcus Porcius Cato the Elder is reported to have been enthralled with the magical aspects of cabbage. When Cato the Elder's wife and son became ill with a fever, he restricted their diet to his beloved vegetable. Despite the fact that efforts here were in vain, he was not deterred from continuing with his own cabbage spree.

In 1689 a prominent Italian physician identified walnut juice as the key to perfect nutrition, a source of long life and health. In 1715, an English surgeon claimed that sugar alone could provide the basis for good nutrition, good disposition, and a cure for all wounds. Another eighteenth-century doctor recommended vinegar as the cure for yellow fever.

Another large group of would-be therapists based their advice on the belief that all plants and animals reflect in their outward characteristics the nature of their curative potential. Thus the lungs of the long-winded fox were recommended for asthma, the juice of a red beet for anemia, and the root of the mandrake for male impotence. The forked appearance of this latter plant was thought to bear resemblance to the two legs of a man. Actually, ingestion of the plant does lend a narcotic effect, so perhaps it did contribute to the composure of the nervous lover.

GRAHAM CRACKERS, CORN FLAKES, AND THE RETURN TO THE GARDEN OF EDEN

It is easy to lose historical perspective for a moment and assume that the back-to-nature mania is an original product of the late 1960's and the early 1970's. But of course "eating natural" is not a new dietary phenomenon. It has just recently resurfaced.

Advances in industrialization seem to bring with them a sentimental enthusiasm for a return to earlier days. Jean Jacques Rousseau (1712–78) set the stage for the Enlightenment with his comments about how civilization made men evil, and why life "in the state of nature" was far superior to forms of sophisticated living. A Swedish contemporary of Rousseau, Professor Carl von Linné (1708–78), was specific in regard to the nutritional aspects of natural living: he recommended that man eat like his nearest cousin, the ape, concentrating on vegetables (especially those available for direct use), milk, fruit, roots, and berries. Linné noted that Swedish peasants had better health than did members of the aristocracy and he linked this observation with the peasants' more basic, less adulterated diet.

Over the past two centuries, numerous vocal nutrition buffs have promoted their own individualized recipes for health, vigor, and longevity. Much of their advice was of the back-to-nature variety. One enthusiast woefully exclaimed: "What a myriad of evils would have been avoided if the first of all food and drug laws had not been violated."

Dr. Sylvester Graham (1794–1851) was the type of

31

food cultist who mixed religious fanaticism with a zeal for the natural, "uncomplicated" life. A member of the Pennsylvania Temperance Society, Graham's primary hypothesis was that man could not survive on sophisticated foods:

> The simpler, plainer and more natural the food of Man is . . . the more healthy, vigorous and long lived will be the body, the more perfect will be all the senses, the more active and powerful may the intellectual and moral facilities be rendered by suitable cultivation.

Graham felt that a truly healthy diet was one that included only what was available in the Garden of Eden: "fruits, nuts, farinaceous seeds and roots," possibly, on occasion, supplemented by a bit of honey and milk. The ideal eating style precluded any form of artificial preparation, except cracking the shells of nuts.

Graham's list of prohibited foods was very long. It excluded salt, or any other condiments (they, like sexual excess, caused insanity), cooked vegetables (against God's laws), tea (it caused delirium tremens), chicken pies (a source of cholera), and liquor (no explanation necessary). But his most forceful and successful efforts were those that focused on the avoidance of blatantly "unnatural" substances and practices such as meat, white flour products, and water consumption during a meal.

His proscriptions against meat were punctuated by a series of less than delightful analogies, including one which claimed that:

> If two healthy, robust men of the same age—the one subsisting principally on flesh meat and the other exclusively on vegetable food and water—be suddenly shot down and killed, in warm weather, and

both bodies be laid out in the ordinary manner, and left to the action of the elements and affinities of the inorganic kingdom, the body of the vegetable eater will remain two to three times as long as the body of the flesh eater will, without becoming intolerably offensive from the processes of putrefication.

As a result of his warnings, a significant number of well-meaning American citizens, not wanting to putrefy prematurely, omitted meat from their diet.

The Graham abhorrence of white flour products precipitated his entry into the entrepreneurial arena. Since it was unnatural to "put asunder that which God hath joined' together," he began to market his own unbolted Graham flour, Graham bread, and those still famous Graham crackers, all of which were exalted in the *Graham Journal of Health and Longevity*. For best effects, the Graham products were to be eaten throughout the day and under no condition was water to be included with any meal. (Anyone who has ever tried eating four consecutive Graham crackers without benefit of some refreshing libation will gain some idea of the degree of loyalty that characterized Graham's followers.)

Graham's galloping enthusiasm, his tendency to weaken his arguments by perpetually overstating them, and the observation that the topics of his supporting material began to become too delicate for mixed company, led to his downfall. He shifted from the health food industry to writing poetry. He died at age fifty-seven, after "stimulants, a tepid bath, and a dose of congress waters."

His ideas, however, lived on. One man influenced by the Graham doctrine was John Harvey Kellogg. While in medical school Dr. Kellogg had lived primarily on a

"Excuse me—I know what I want, and I want what I asked for— TOASTED CORN FLAKES —Good day"

he package of the genuine bears this signature
W. K. Kellogg

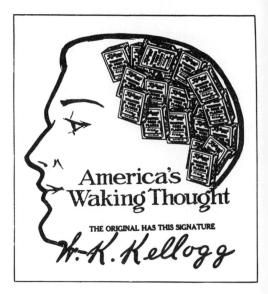

America's Waking Thought

THE ORIGINAL HAS THIS SIGNATURE

W. K. Kellogg

NE GENUINE WITHOUT THIS SIGNATURE
W. K. Kellogg

"I Love
my Jam—
But
O You
Toasted
Corn
Flakes"

A TREAT FOR ANY MEAL

KELLOGG'S Corn Flakes are delicious for breakfast, lunch, late snacks, children's suppers. Healthful, easy to digest — wonderfully appetizing.

And think of all the work and money you save. All prepared — trouble-free. Enjoy with milk or cream, fruits or honey added.

You get many servings from a single package costing but a few cents. Always oven-fresh in the patented sealed WAXTITE inside bag. Do you wonder that Kellogg's Corn Flakes are the most popular ready-to-eat cereal? Sold by all grocers. Made by Kellogg in Battle Creek. Quality guaranteed.

LORETTA YOUNG — First National Pictures star

Enjoy a bowl of Kellogg's

Advt. No. 33814

In two jiffies a flavory meal to satisfy the hungriest man —Kellogg's Crispy Corn Flakes.

1st It's a meal to make a hungry man beam with joy. With milk or cream, as satisfying as it is tasty.

2nd Good for every man who uses up a lot of body-fuel. Builds energy fast.

3rd It is ready to eat. No cooking. No waiting. Just pour in a bowl and serve.

Get the genuine Kellogg flavor by asking in full for Kellogg's Corn Flakes.

Kellogg's CORN FLAKES

New York Sign Completed

The Kellogg sign at the Times Square on the roof of the Mecca Building, has been completed and is now in operation. It holds the distinction of being the largest electric sign ever built.

FRESHER THAN EVER BEFORE..!

EXCLUSIVE NEW "WEATHER-PROOF WRAP"

NOW BRINGS YOU AN ENTIRELY NEW FLAVOR SENSATION IN KELLOGG'S CORN FLAKES!

This pig-tailed young lady is a small neighbor friend of Norman Rockwell up in Arlington, Vermont.

Mr. Rockwell has painted her many times before, but never eating Kellogg's Corn Flakes.

She eats them all the time, however, and this is how she looks. She likes them, as you can plainly see. So does Mr. Rockwell. So do most people.

These are the friendly flakes that get a real hold on your appetite. Small moppets have been stowing them away and spilling a few in their bathrobe pockets for over 50 years. They're the cereal that's most often back of father's newspaper . . . the cereal mother likes, too, when she has time to eat . . . the cereal that millions have grown up with and never said goodbye to.

The problem, of course, is that people are always running out of Kellogg's Corn Flakes and that's a very good way to make someone cross at breakfast. How is it with you? Need another package?

Courtesy of Kellogg Company

diet of apples and Graham crackers (his normal breakfast was reported to be seven crackers, two apples—and no liquid).

Dr. Kellogg was a member of a Seventh Day Adventist group which, around the turn of the century, formed a religious colony and health sanitarium at Battle Creek, Michigan. It is said that Dr. Kellogg and his brother, Will, were the first men to make a million dollars from food faddism.

The Kelloggs believed that it was God's law that man eat only plants and fruit. Dr. Kellogg even imposed these restrictions on his dog, Duke, although it is rumored that, quite furtively, Duke was on friendly terms with the local butcher.

In treating the patients who came to his sanitarium, Dr. Kellogg would serve them granula (a predecessor of our granola that was a leftover bread that had been dried in the oven and then ground up) and zwieback, a raised bread that was sliced and then overbaked. But this diet, no matter how healthy and religiously correct it was, did present problems. For instance, one sanitarium patient broke his tooth and demanded a very expensive replacement.

The Kellogg brothers decided that there had to be a better way. Research efforts intensified and soon "wheat flakes" made their debut.

By 1899, Will Kellogg had built up an impressive corporation and the cereal-based health food industry was well established. Competitors began to sprout up all over. One of those looking for a piece of what appeared to be a very promising market was a former patient at the Kellogg sanitarium, Charles W. Post. He invented another type of ready-to-eat "health cereal" by baking wheat and barley loaves, then drying and grinding them into small, gravellike bits. He called them grape nuts (mysteriously, since neither grapes nor nuts have any part in their makeup) and marketed them as a cure for appendicitis, loose teeth, consumption, and malaria. Mr. Post always included a copy of his booklet "The Road to Wellville" in each package of his product.

From these beginnings with cereal health foods developed one of the large and outstanding segments of our modern food industry, namely the breakfast food industry as represented by the Kellogg Company, the Post Division of General Foods, and several other companies with excellent research, marketing, and public service activities.

Graham, Kellogg, and Post appear to have been relatively rational compared to some of the health faddists who followed them. Adolphus Hohensee, for instance, in the 1940's headed up the type of health movement that led some people to wonder if they dared to sit down to dinner again.

THE WORMS WILL GET YOU IF YOU DON'T WATCH OUT

Adolphus Hohensee, whom we've already mentioned in connection with the alleged cleansing potential of garlic, began his career in the field of nutrition as a soda jerk. He then moved into the real estate business and met with a great deal of success until he was jailed for mail fraud and passing bad checks. His quest to redeem and reorient himself led to his application for, and eventual receipt of, an Honorary Doctor of Medicine degree from the unaccredited Kansas City University School of Physicians and Surgeons. He received this degree in 1943, which was fortuitous timing in that the institution was closed down in 1944.

"Dr." Hohensee decided that the health food business was for him. He immediately selected an effective and very traditional approach to the ears, minds, and stomachs of unsuspecting consumers: scare tactics to create a concern about the safety of regularly available food, followed by a spectacular presentation of his own products which were guaranteed to meet every need.

He apparently enjoyed the showmanship approach. He often began by telling his audience that 90 percent of them had worms. Then he vividly described the worms. They were two to twenty feet long. They had their heads in the stomach of their victim, and their squirmy bodies extended into the intestines.

But all was not lost. The Hohensee special cleansing diet could make you worm-free.

Then he moved to raise anxieties even further by announcing that Americans, as a result of their diet, were generally suffering from poor health. In addition to having parasites, they were unable to perform sexually.

According to Hohensee doctrine, the sex act itself, apart from any preliminaries, was to last at least one full hour. (Those whom he didn't get with the worm scare, he got here.) If you did not measure up to this hallmark of sexual adequacy, it was probably because you had the wrong diet. The average American diet stagnated the blood, corroded the blood vessels, eroded the kidneys, and clogged the intestines. And food additives, according to this twentieth-century food prophet, were just the worst of all dietary scoundrels, causing the type of slow poisoning process that inevitably followed the consumption of "dead" processed foods. Hohensee claimed that additives were being used because of the interests of the American Murder Association, whose principal members were the Food and Drug Administration, the Better Business Bureau, and the food manufacturers.

To ensure a long life, according to Hohensee (to "live to be 180 . . . and go to the moon"), one had to follow a three-phase approach: the daily consumption of Hohensee-endorsed wonder products (which happened to be well-stocked in the health food store adjacent to the lecture hall), the avoidance of certain disease-inducing foods (liquor, white bread, and particularly fried foods), and the regular use of garlic suppositories.

The FDA and other regulatory agencies got after Hohensee on a few occasions and eventually succeeded in fining him for misrepresenting his products. But what really led to his downfall was not those relatively minor legal actions, but rather the type of event about which food-fad critics wistfully dream.

After a particularly successful and economically rewarding week of lecturing, Hohensee decided to celebrate with a night out on the town. He quietly slipped into a restaurant in Houston and, thinking he was alone and unnoticed, ordered a canapé of forbidden foods including fried snapper (from the "death-dealing skillet") and thick slices of white French bread ("it knots in a ball in your stomach, and stays there in a big lump"), which he washed down with big gulps of cold, foamy beer (he condemned all booze and "barflies"). Unfortunately for Hohensee, and fortunately for the people he was duping out of thousands of dollars, the *Houston Press* and other papers discovered him *in medias res*, and subsequently published headlines that read "What's Up, Doc?" and " 'Nature Doc' Dines Out and Knocks a Decade off His 180-Year Life Span."

Faddists Galore

There were other would-be nutritional reformists. For instance, Gaylord Hauser, author of *Look Younger, Live Longer* (which was on the nonfiction best-seller list in 1950 and led the list in 1951), "discovered" that the miracle of life lay in the naturopathic approach.

With Greta Garbo as one of his major enthusiasts, Hauser recommended five wonder foods: skim milk, brewer's yeast, wheat germ, yogurt, and blackstrap molasses. Eating a diet that revolved around these foods would, he claimed, add years to your life. Evidently people believed him too. He became a popular health lecturer and managed to sell over a third of a million copies of his book. Some of his later works, including *The New Diet Does It* ("revised and enlarged") and *Mirror, Mirror on the Wall,* are currently being promoted in health food stores.

Of course, Hauser did have some critics. Jimmy Durante once argued that these special health foods did not really make you live longer; they just made it seem longer.

At the time of this writing, Gayelord Hauser is eighty years old (so are a few hundred thousand other Americans who presumably did not live on Hauser wonder foods) and he's still collecting royalties and promoting his "million-dollar secrets for health, beauty, and *joie de vivre*." His latest book, *New Treasury of Secrets,* includes descriptions of "delicious wonder foods that quickly help to melt your fat—and conquer your craving to overeat," "the incredible Swiss apple diet that lowers blood pressure and reduces inches," "complete directions for growing your own precious vitamins," information on "the super-energy eye-opener, drink it the minute you jump out of bed," and "how Gayelord Hauser saved the world's most glamorous women from mid-afternoon slump."

There was Dr. William Howard Hay (a medical doctor), whose book *Health via Food* (1933) was received with much acclaim by food faddists. He felt that American problems of all kinds were related to three dietary shortcomings: eating too much protein, focusing on "adulterated" food like white bread, and retaining food in the bowels for more than twenty-four hours after eating. He recommended changing the diet to reduce its protein content, eating "natural" foods, fasting, and frequent use of laxatives. He also offered some alternatives to regular foods, including Parcel Post Asparagus, Easter Bunny Salad, Hay's Happy Highball, and Pale Moon Cocktail. Apparently Dr. Hay did not share his predecessors' aversion to the fruit of the vine.

In 1958, Dr. D. C. Jarvis (another medical doctor, who was a member of the American Medical Association's ophthalmology and otology sections) made another contribution to the growing library of food faddism. His *Folk Medicine: A Vermont Doctor's Guide to Good Health* revived an eighteenth-century interest in vinegar.

According to Dr. Jarvis, the principal American health threat was an excess of bodily alkalinity. To increase the acid component and create an environment that would prevent "the infestation of the body with pathogenic microorganisms," patients were advised to avoid meat, wheat foods, citrus fruits, white sugar, and maple sugar. In addition, if patients would precede each meal with two tablespoons of apple cider vinegar, acidity allegedly would be increased.

A daily urine check was advised to measure progress in the fight against alkalinity.

Professional colleagues who reviewed Dr. Jarvis' book fervently hoped that he was a better ophthalmologist than he was a nutritionist.

SUPER-NATURAL: FAMOUS FADDISTS OF THE 1960's AND 1970's

Jerome Irving Rodale, the son of a grocer and the founding father of our twentieth-century Healthfoodland, began his business career as a manufacturer of electrical wiring accessories. Very soon, however, he discovered a far more rewarding economic outlet and made an official vocational switch to become the principal promoter of natural, organic living.

Rodale, following the dogma of British agronomist Sir Albert Howard, maintained that in order to remain healthy and ensure a long life we had to eat "God's way." First, according to the Rodale Doctrine, we should use only natural, organic matter in fertilizing our soil. Unlike the prepared chemical fertilizer, organic material allegedly guaranteed that the soil would hold moisture and retain its "tillability." Furthermore, the organic farming methods, as opposed to "artificial means," were said to provide a more basic, balanced form of nutrient for plants.

Second, foods were to be eaten directly, with no artificial anything and without benefit of cooking. Rodale always solemnly pointed out that "no animal eats cooked food." And since none of his critics had ever seen a cow with a frying pan, they had to agree with this.

Third, it was vital that certain foods be avoided. Wheat, for instance, was on the forbidden list since, according to Rodale, it made you overaggressive and "daffy." Sugar was even worse and was proscribed for all organic eaters—and fluoridated water was to be avoided even if it meant buying bottled water.

Fourth, it was essential that the organic, "chemical-free" diet be supplemented daily by rose hips, natural vitamins and minerals, and other such health enhancers. Mr. Rodale and his wife took seventy different types of supplements each day.

Jerome Rodale's nutritional advice was promoted both through his books (including *Organic Front* and *Happy*

People Rarely Get Cancer) and in the highly successful magazines and newsletters that regularly rolled off the Rodale Press—and still do: *Prevention* (launched in 1950, now claiming an audience of over one million), *Organic Gardening and Farming* (begun in 1942, with a current circulation of about 700,000), and *Fitness for Living.*

The fact that the American Medical Association includes the Rodale publications in its quackery exhibits apparently has not had a significantly depressing effect on business. Readers seem willing to believe his medical advice (including the recommendation to eat pumpkin seeds to cure prostatic disease) and in 1970 alone, Rodale and his associates reaped some nine million dollars from their advocacy of natural living.

Jerome Rodale insisted that his healthy life-style would allow him to live to be a hundred years old ("unless I'm run down by a sugar-crazed taxi driver"), but he didn't make it to the century mark. He died at the very respectable age of seventy-three after being stricken while taping an interview for the Dick Cavett show.) His son, Robert Rodale, has inherited the health food leadership and, among other things, has assumed the editorship of *Prevention.*

Adelle Davis was the guru of lay nutritionists. According to her philosophy, proper diet could not only prevent and cure any disease, but could have important effects on the quality of life in society as a whole.

Actually, Mrs. Davis did have some educational background in the area of nutrition. She was trained in dietetics and nutrition at the University of California at Berkeley and received an M.S. in biochemistry from the University of Southern California Medical School in 1938. Her former classmates and teachers in these programs are reported to have affectionate memories of her and her potential for improving the levels of public nutrition knowledge—and they were distressed and disappointed at the direction her efforts eventually took.

What foods did Adelle Davis recommend avoiding? At the top of the "not to be eaten or drunk" list were refined sugar, pasteurized or homogenized milk, white bread, food additives, "unfertile" eggs, and any food that may have come in contact with "chemical fertilizers." For Mrs. Davis "proper diet" meant focusing on whole grains, fresh milk, fruits and vegetables, and at least two (fertile) eggs and a helping of cheese every day (she was not in agreement with the prevailing medical opinion linking foods high in saturated fats with heart disease).

She wrote articles and books (including *Let's Eat Right to Keep Fit*, *Let's Cook It Right*, *Let's Get Well*, *Let's Have Healthy Children*), lectured and made television appearances, all in an effort to get across her line about the critical role a healthy diet can play. "A woman," she gravely warned, "who wants to murder her husband can do it through the kitchen. There won't even be an inquest." And she offered thousands of respectable-sounding references to back up her statements and advice (in *Let's Get Well*, she lists 2,402 references to document thirty-four chapters). At first glance her work might appear impressively well grounded, but when Dr. Edward H. Rynearson, professor emeritus of medicine at the Mayo Clinic and Mayo Foundation, actually attempted to track down the "references" for a number of her statements, he found that a significant number of her comments had no basis at all. In Chapter Twelve of *Let's Get Well*, for instance, he noted that of the fifty-seven references listed, twenty-seven had no data supporting her statements. Dr. George Mann of the

Department of Biochemistry and Medicine at Vanderbilt University School of Medicine read her most popular book, *Let's Eat Right to Keep Fit,* and found, on the average, one mistake per page.

Some of her misleading advice caused concern in the medical profession. For instance, when she took a stance against bottle-feeding and stated that crib deaths could be prevented by breast-feeding and diet supplements of vitamin E, at least one physician responded by sending a letter of complaint to the Federal Communications Commission asking for action against these statements. (Mrs. Davis replied to this physician's inquiry by writing "Thank you so very much for correcting me. It was the first time I had heard that crib death occurs in infants while they are being breast-fed. I am indeed sorry if words of mine have added to the suffering of parents whose infants have died." The problem with Mrs. Davis' apologies was that they never got the attention her original inaccurate statements did.) In *Let's Get Well,* Mrs. Davis recommended that patients with nephrosis take potassium chloride, a suggestion that according to medical specialists is "extremely dangerous and even potentially lethal."

Some of the historical and sociological evidence Mrs. Davis offered to defend her diet theory raised many eyebrows and must have stimulated a few disbelieving chuckles among her critics. For instance, she pointed out that the reason the Germans were able to overpower the French during World War II was definitely diet-related: black bread and beer, she felt, were nutritionally superior to white bread and wine. Similarly, when a group of mass murderers was arrested in Southern California, Mrs. Davis explained that their behavior was the direct result of their exclusive diet of candy bars.

45

At age sixty-nine, Adelle Davis was informed that she had bone cancer. She was initially shocked that,' despite all her efforts to ensure healthful living, she could be affected by such a disease. But she told her followers that, although she had eaten well on the Indiana farm where she had been raised, she had reverted to "junk" foods when she went away to college, and remained with a nutritionally inadequate diet until well into the 1950's. Just before her death, about a year after her illness was diagnosed, Mrs. Davis fervently urged her followers not to be disheartened, but to continue to seek long life and good health through "natural" products.

FOOD FADS IN PERSPECTIVE

There have always been food fads, many of them based on the return-to-nature theme. Our current preoccupation with "100% natural" products, a fad based largely on the philosophies of Rodale and Davis, is much like the food binges of the past in that it appears to provide a way of meeting certain deep-seated emotional needs. But it also has some other aspects similar to food manias of other times.

Food and eating habits offer a type of bond that can bring people together and reinforce their unity. Fads, therefore, can serve the purpose of identifying some specific subculture.

All major religions, for instance, have some form of food proscription which may uniquely characterize their particular sect. Pope Gregory III ordered his apostle Boniface to forbid the consumption of horse flesh by all Christian converts as a means of establishing a sense of separateness from the pagan tribes who ate it and made horse meat a part of their rites of worship. Buddhists believe "not to injure living things is good," thus animals may not be slaughtered and used for food. Jews do not include pork or shellfish in their diet, avoid the combination of meat and dairy products, and maintain separate dishes for these courses. Catholics periodically restrict the use of meat.

And this type of "food separatism" goes beyond religious boundaries. Today, upper-class businessmen are uniting in a concern over cholesterol (a legitimate concern), lower- and middle-income working men are focusing on getting as much red beef as they can afford, and those members of our society who have an intense disdain for big institutions and the products of technology are eating "100% natural food." It is this suspicion, the negative feelings about the military-medical-industrial complex, and the associated nostalgia for "plain living" that provide the bond uniting many (but certainly not all) of today's natural-food faddists.

One Berkeley, California food store owner put it this way: "I see [natural food] as a way of understanding oppressed people and the political implications of food. Eating natural food is a way of subverting Safeway. The food industry is a heavy political thing."

The "us versus them" attitude expressed by this Berkeley natural eater is an example of a broad feeling of suspicion the "consumer-environmentalists" have for the "greedy industrialists." It all started with the publication of Rachel Carson's *Silent Spring* and the identification of these two allegedly opposing interest groups.

The food-scare books of the late 1960's and early 1970's reinforced this dichotomy, claiming that food manufacturers were in collusion with the FDA to cheat the unsuspecting consumers:

47

Diethylstilbestrol [DES] is a synthetic estrogen . . .
used in cattle raising. The sole purpose in slipping
it to animals, as is the case with almost everything
we'll talk about that goes into the food you eat, is
so that someone can make an extra buck. . . .*

Apparently, people expressing this point of view are un-
aware of the theory of supply and demand, and choose
to overlook the fact that a measure taken to increase
the supply of any product will be reflected in lower
prices. All they perceive is an "us vs. them" attitude.

To ensure that the economic returns in the food-fad
business are large and continuous, the faddist-promoter
has traditionally turned to an old and very successful
technique: spreading rumors and raising suspicion about
the opposition, in this case, regular food. The nineteenth-
century natural food lecturers did exactly this. And just
a few years ago the makers of iron pots "informed" the
public that aluminum causes cancer. There has always
been economic advantage in undermining the competition with a series of downright false statements carefully mixed with a few half truths for good measure.
And, as we can observe today, things don't change much.
"Natural food" promoters continuously claim that foods
with additives are health-threatening, and they strongly
suggest that "unnatural" eating is a primary cause of
cancer.

And the health food route continues to ensure substantial profits for those who succeed in taking advantage
of deep-seated fears about illness. Groucho Marx was
once interviewing a health food promoter on his radio
show and asked the faddist what his product was good

* Marine, G. and J. Van Allen. *Food Pollution: The Violation of
Our Inner Ecology*, New York: Holt, Rinehart & Winston, 1972.

for. With unusual candor the health promoter replied, "It was good for about five and a half million last year."

So food fads are nothing new. But what are some of the characteristics of today's preoccupation with natural products? How is it different from past fads? What types of products are being promoted? And what are some of the problems being raised?

THREE

The Garden of Eden:
1970's Style

"There are more fools than knaves in the
world, else the knaves would not have
enough to live upon."

SAMUEL BUTLER, Notebooks

THERE'S MONEY IN HONEY!

HISTORY is full of food fads. But our current preoccupation with organic and additive-free food has assumed truly unprecedented dimensions. Sylvester Graham would have been pleased—and probably a bit envious.

When Graham opened the first American health food store in the 1830's, he was pretty much a loner. His clientele was primarily limited to the small portion of the population that attended his lectures, subscribed to his journal, or followed up the advertisements he placed in the local newspapers. The same was true of the brigade of food faddists who followed him, lecturing

50

their way through the next century. The early health food businesses catered to the emotional needs of a relatively small portion of American eaters. Even Hohensee, during the 1940's and 1950's, and Dr. Jarvis in the late 1950's and early 1960's generated only a moderate amount of enthusiasm.

But the health food phenomenon we are observing today is different. It's *really* big business. Health foods have made the transition from being the focus of a fad to becoming the material of a movement.

The *Wall Street Journal* estimated that in 1970–1971 there were 1,500–2,000 individual health food stores in operation in the United States. Others have estimated that by 1972, the total reached well over 3,000. In New York City alone, there were, at last count, over 175 outlets that featured the natural life—and the growth trend shows no sign of leveling off.

In addition to the health food outlets, there is a significant number of restaurants that specialize in allegedly health-giving delicacies. All told, it is estimated that Mother Nature and her earthly associates are benefiting from a 100% natural business yielding one billion dollars a year. It is projected that retail sales by 1980 will be at least three billion dollars.

There was nothing gradual about the arrival of the modern health food movement. It struck like lightning during the early part of 1970. During the 1960's, of course, particularly after the 1962 release of Rachel Carson's *Silent Spring*, there was new concern about various types of environmental contamination. But it was the 1970–72 period that saw the unprecedented increase in the health food industry.

By 1971, the majority of students at the University of California at Santa Cruz stated that they preferred

A MODERN FAIRY TALE

ARTIST & WRITER: SERGIO ARAGONES

NO PRESERVATIVES ADDED
NO ARTIFICIAL COLORING
100% PURE ORGANIC
NO CYCLAMATES

natural and organic foods over the more traditional variety. By 1972, it was reported that crunchy granola (an oat flake cereal, sometimes coated with honey) was doubling in sales every two months.

THE BILL OF FARE

To be a natural organic shopper, and to rationalize spending 50 percent more for so-called health foods, you've really got to believe. Generally, the creed of natural organic eating is based on four main premises. These premises need not have any factual basis. They claim the same status as a code of religious dogma.

The four cornerstones of Healthfoodland are:

Organically grown food is more nutritious and is safer.
"Natural" is better than "artificial."
Some foods have "magical" properties.
Vitamins and other food supplements are wonder capsules, and you can't have enough of them.

The number of products that seem to meet these qualifications is relatively small (compared to what is available in more traditional food stores), but today's Garden of Eden does offer considerably more than Graham crackers and garlic. The natural selection now covers the gamut from asparagus to zwieback—and in between includes cosmetics, cooking apparatus, and liquor.

A faddist entering a health food store might first be confronted with a choice of milk: the cow variety and the tiger variety. If he chooses cow's milk, he is advised to drink it in its original form, that is, unpasteurized and freshly released from the beast. Ideally, according to one natural food enthusiast, he should have his own cow. But if he lives in an apartment where grazing would prove difficult, the products available at the local health food spa will do just fine.

But then there's Tiger's Milk, which is not, as the name implies, a feline secretion. Rather, to quote the label, it is a concoction of nonfat dry milk, raw sugar, yeast, sodium cassinate, tricalcium phosphate, emulsifier, and natural and artificial flavor. (What? Strange-sounding chemicals in a health food?) Basically Tiger's Milk is usually a version of yesterday's Hauser-miracle-food diet. The missing items—for instance, wheat germ and honey (although sometimes they are included)—are part of any loyal faddist's daily intake anyway.

The choice of cereals is even larger. There are the hugely successful crunchy granola with many variations, wheat germ in all its varieties, vitamin-enriched grains, nuts, raisins, and (if you can overlook its vaguely canine appellation) crispy Alpen, which is "made with good things from the good earth."

If natural-oriented consumers have nothing else to do all day, they can occupy themselves by buying and using the ingredients for homemade bread. Of course, this will have to be an almost daily endeavor, in that bread without preservatives tends to lose its freshness with amazing alacrity. But then who minds a mere ten to fifteen hours a week of breadmaking-for-health?

The dedicated faddists will also have to spend some portion of their week carefully shopping for meats from animals that were grown without benefit of "hormones," and poultry that was allowed to "roam free with the rooster." They also must find fish caught in the "pollution-free Atlantic Ocean," and organic eggs which were laid by happy hens. On busy days, however, they can choose from a wide selection of frozen natural goodies, including organic stuffed cabbage, chicken casserole with natural herbs and brown rice, and good old-fashioned soybean sprouts.

But naturalists cannot rely just on food—even if it is 100% organic and additive-free. They need vitamins, minerals, protein supplements, desiccated liver, wheat germ oil, and a whole host of allegedly life-giving extras. And it's not just human consumers who need these products. Cats and dogs do also. After all, you have to protect *their* nutritional well-being, too.

And there is more. To be truly healthy, the food faddist must clean himself out. Seawater is very useful for this purpose. But natural laxatives are also available. Some health food stores feature over thirty-one different type of laxatives (generally made of papain, whey, and alfalfa), some of which have been advertised as means of "cultivating your inner organic garden."

Garlic tablets and a variety of other garlic products, including garlic oil ("for healthier arteries") and croutons and crackers, are recommended. No direct mention is made of the time-honored interpretation of the alleged purgative effects of garlic. And no advice (or reference to Hohensee counsel) is offered with respect to how the garlic product is best used. Perhaps it is assumed that any health food faddist worth his sodium chloride will innately know what to do with it.

Industrial Giants and Naturalism

As pointed out by Dr. Virgil O. Wodicka, then director of the FDA's Bureau of Foods, "This back-to-nature bit is causing some of the traditional food industries to try to take some advantage of the nomenclature." Going natural is by no means limited to the local health mini-entrepreneurs. Indeed there is evidence that food companies are finding it more expedient and profitable to get on the back-to-nature bandwagon and exploit the natural food myths rather than sponsor an educational campaign to fight food faddism.

Coca-Cola brags of "the real thing." Jones Dairy Farm offers sausage without "unnatural" preservatives. Dannon boasts of yogurt with no chemical additives ("it's only natural"). Certain beverage companies (the ones who "know how you feel about beer") have "gone natural." And the butter industry is now attempting a comeback, advertising its product as "a new margarine substitute—free from chemical additives, based on an old family recipe passed down from cow to cow."

According to the *New York Times*, the hottest category in the breakfast cereal business these days is natural or health cereal—direct descendants of granola. Quaker Oats has Quaker 100% Natural Cereal ("no synthetics; no mystery ingredients"), Pet has Heartland, General Mills promotes its Nature Valley, and Kellogg has a product called Country Valley. When the Quaker Oats 100% Natural product was on the market for just a few weeks, William D. Smithbury, a Quaker representative, reported, "It was obvious we were riding a tiger." He further noted that advertising had to be cut back because they could not keep up with the demand.

A drawback of these so-called natural cereals—both those produced by the traditional food industries and those of the marginal food operations—is that they are simply loaded with honey, sugar, and fat. The result is a high calorie product of up to 140 calories per ounce (as opposed, for instance to the 90–100 calories per ounce content of the ordinary prepared cereals).

As well as becoming naturally nourished, the health enthusiast can also become naturally fat.

FDA officials are expressing some concern about this industrial trend toward "naturalness," since there are potential implications that are far more serious than extra calories. If an independent meat company decides all of a sudden to "go natural" and leave nitrates and nitrites out of its products, will officials have to wait for the first case of botulism to occur before they step in to stop the practice?

NATURAL VICES

Unbelievable though it is, at least one cigarette manufacturer has attempted to get on the back-to-nature bandwagon. How any product that carries the grim warning that cigarette packages must can be so brazen as to have a sybaritic lady sitting by a trickling brook extolling the virtues of "naturally grown menthol, rich natural tobacco taste . . ." is not fully clear. But it is done.

Even more inconsistent is a popular product offered at many health food stores—a form of natural herb candy to relieve the scratchy irritations of smoker's cough.

A health food store entrepreneur was recently observed talking up the superior attributes of his products, and the health-threatening aspects of the foods in the store adjacent to his. He said, "Why, you don't know what you're eating these days. All those complicated-sounding chemicals you see on packaging labels. They could cause cancer for all you know!" He then excused himself to find an ashtray to accommodate the precariously balanced ashes on his smoldering weed.

Liquor is in the act too, and there are claims of a "natural booze."

Dr. Frederic Damrau and Professor Arthur H. Goldberg at Columbia University's School of Pharmaceutical Sciences confirmed that it is the "additives" or congeners (the esters and other impurities in some alcoholic beverages that give liquor its flavor and bouquet), not the booze itself, which cause the morning-after syndrome. And immediately afterward, there was a rush on "natural" cocktails, that is, those made with vodka.

Vodka is filtered through charcoal, a recognized drug listed in the United States Pharmacopaeia and a standard antidote for many poisons, sold today at drugstores as a treatment for gastric distress due to overeating. The use of charcoal dates back to the time of Hippocrates, but only in the latter part of the eighteenth century did its absorbent action become clear and just recently it was reported that activated charcoal could remove the "hangover material" from standard-brand 86-proof whiskey.

Tests with volunteers have confirmed that subjects drinking vodka, which is alcohol reduced to its natural—and tasteless—state, demonstrated significantly fewer hangovers (including gastric irritation, headaches, dizziness, fatigue) as compared to those drinking non-charcoal-filtered alcohol. When volunteers were given "zero proof" whiskey, that is, liquid made by removing all the alcohol and leaving only water and congeners, as little as one ounce caused hangover symptoms in more than half the subjects.

A number of medical researchers have confirmed this observation and one London physician is on record as stating that "the aftereffects of vodka are substantially less than the equivalent amount of any other kind of alcoholic drink."

Apparently, the Russians figured that out years ago. But the problem in this country is that many people like the congeners which give the whiskey its taste, bouquet, and color and do not choose to serve charcoal bits with their pretzels and peanuts to conduct the filtering process on an internal basis.

NATURAL LOOKING AND NATURAL COOKING

If you treat the inside of you to natural goodness, why not the outside of you too? So goes the reasoning of the natural cosmetic industry.

If you take a tour of the cosmetics section of any of the big department stores, you'll find a rather elaborate selection of natural health care products, most of which sound good enough to eat.

There's a creamy milk-bath selection; honey shampoo; wheat germ and yeast beauty masks; vitamin hair tonic; cucumber, coconut oil, tomato or "milk and honee" soap—and for dessert, a choice of strawberry, peach, or apple cleansers.

Look young. Go natural. You can wash your hair with organic Herbal Essence shampoo and "launch on a totally new experience" with Elizabeth Arden's multiple-action cleansing cream ("suddenly your face feels invigorated with a look of natural health"). Don't pollute your skin with suspicious artificial goo. According to one promoter, using regular cosmetics is like eating white bread with preservatives added.

And if you're going natural, you might as well go all the way. At least that's what the manufacturers of "total organic environment" products hope you'll do. You'll certainly want to throw out all your aluminum cooking ware, and replace it with a set of stainless steel ware which is "guaranteed" not to contribute any metal materials to your organic gourmet delights. And you'll need a blender for making your baby's food, biodegradable paper towels, organic toothpaste, a yogurt maker, a water-purifying system, and a grow-your-own-herbs-at-home kit. That's just for starters.

THE BOTTOM LINE

Before you decide to go natural, you'd better check your bank balance. The markup on miracle foods is *truly* healthy.

The difference in price is related first to the fact that merchants are reaping the usual type of economic benefit that goes with dealing in specialty foods. But second, it is directly related to the fact that the production of food grown in "organic" soil and raised without benefit of pesticides inherently costs more. For instance, while the nitrogen supplied by commercial fertilizer costs between 7½ cents and 15 cents per pound, an organic farmer will pay $12 or more for the same pound of nitrogen derived from garbage compost or $5 a pound for dried cow manure.

It's really difficult to say how much you will be taken for when you take your first step into Healthfoodland. If you buy a big bottle of seawater, having been convinced that "from the sea comes life," you are really being soaked. As a matter of fact, you might as well have purchased the Brooklyn Bridge. At least you wouldn't have to carry that home.

If you purchase solar-evaporated sea salt (which carries a label "this salt does not supply iodide, a necessary nutrient"), you've really been doubly sopped, since you will then have to purchase iodide tablets to supplement the salt you just paid more for because it was iodide-free.

The markup of more normal food varies from place to place. An informal survey of stores in the New York City area yielded this comparison:

	REGULAR STORE	HEALTHFOODLAND
Canned apple juice (qt.)	.49	.89
Dried pitted peaches	.86	1.68
Cornmeal	.18	.55
Honey (1 lb.)	1.63	2.99
Cucumber	.24	.89
Salt (26 oz.)	.11	.39
Mayonnaise (1 pt.)	.67	1.98
Cooking oil (1 qt. 6 oz.)	1.69	2.15
Ground chuck (1 lb.)	1.29	2.15
Eggs (extra large, dozen)	1.09	1.79

In a decade of unbridled inflation, there must be some pretty spectacular types of incentive to convince people to accept such a significant dollar margin. Generally, natural consumers do believe there are such incentives. One organically satisfied customer explained that the organic food is more nutritious, thus one has to eat less, and it's all balanced out in the long run.

WILL THE REAL ORGANIC CARROT PLEASE STAND UP?

FOOD FRAUD

In both a scientific and dictionary sense of the term, all foods derived from living organisms are organic. Calling a particular type of food "organic" is like labeling a specific form of water "wet."

The combined effects of a lack of universal definitions for "organic" and "natural," the absence of any practical means of identifying foods from organic sources, and the presence of a loyal and unsuspecting health-oriented consumer population contribute to a perfect setting for food fraud.

What does "organically grown" mean? Does it mean

that the grocer whispered "organic" over the vegetables as he carried them to a special, higher-priced shelf? Does it mean that the product was tested to ensure that it had no pesticide residues? The FDA recently discovered a "health food" store in Boston which was ripping off the public by ripping off traditional labels, replacing them with "organic-natural" designations. But usually food frauds are a bit more subtle than this.

As it stands now, the terms organic and natural can mean anything the merchant wants them to mean. Peter Barton Hutt, general counsel of the FDA, summarized the feelings of most observers when he said:

> I don't think there's any doubt that there's a hell of a lot of fraud going on. It's very simple to put a little sign over some raw produce and say "organically grown."
> Who in heaven's name will ever be able to figure out whether it is or it isn't? Who's going to test it, who's going to make sure it's not inaccurately labeled?

It occurred to New York officials that there was big business in misrepresenting "organic" and "natural" foods. And Attorney General Louis J. Lefkowitz decided to look into the matter. Confidential investigators from the Bureau of Consumer Frauds and Protection anonymously shopping in about twenty-five health food stores in New York City first asked the merchants to describe the benefits of their products over more traditional foods. They were told that the health foods offered were pesticide-free, more nutritious, and generally healthier in the long run. Dr. Elmer George, director of the New York State Food Laboratory, then tested about fifty-five "natural," "pesticide-free" items. He found that fully 30 *percent of them had pesticide residues* (as compared to about 20 percent of regular foods). In his analysis of beets, he found .03 ppm of DDT, about the same as regular beets.

Even a farmer on his best organic behavior can end up with pesticides in his food. Residue chemicals can be deposited by the wind and water. They're almost impossible to avoid. There is no way of testing a product to see what kind of fertilizer was used. So you're left with a situation where certain specialty foods are being sold at a significantly higher price, yet there is no easy way to establish if they are frauds.

The real problem is the lack of a standard definition. It's not that people haven't *thought* about a definition: nine members of the faculty of the Department of Nutrition at the Harvard School of Public Health looked very carefully at the question of a definition of "health foods" and came up with the recommendations that (a) *no* foods be identified as health foods because all edible foods, when properly used in a balanced diet, are conducive to physiologic or psychologic health, (b) regulations be adopted to ensure that *no* foods are singled out as "organic" because all foods are organic, and (c) regulations be adopted prohibiting the use of the term "natural foods," because all foods are natural or are manufactured from natural foods.

But various groups have moved in the direction of developing definitions. The Rodale Press has its guidelines. The California legislature has a statute that offers some definitions. And the FDA has drafted its own statements about what is, and what is not, an "organic, natural" food. Of course, all of these definitions are very divergent and according to FDA counsel Hutt, "Most of them, including ours, are sheer nonsense."

MYTHS AND REALITY
IN HEALTHFOODLAND

All types of concerned individuals are now turning to the Mother Nature spas for nourishment and emotional security. It's not just the members of a united anti-industrial "subculture" who are electing to eat natural.

A study conducted by the University of Hawaii concluded that some "reasonably normal people," not just young, long-haired radical members of the lunatic fringe, shop at health food stores, reporting that they get most of their nutritional information from Adelle Davis and the Rodale Press.

Maybe some of these "reasonably normal people" think the food at Mother Nature Corner is fresher. Or perhaps, as a number of nutrition experts have hypothesized, some get some psychological benefit from natural foods—see them as a type of panacea. For some people, natural foods may be a form of tranquilizer without any of the undesirable side effects. But generally, people visit Healthfoodland to buy food they feel is safer, more nutritious, and possibly even capable of providing a magical curative or preventive effect. But let's take a look at some of the health faddists' major premises and the problems of the "eat natural and be sure to supplement" way of life.

MYTH #1: Natural and Organic Is Better

REALITY

First, there is no basis to the belief that organic food is nutritionally superior. An apple is an apple. You simply can't change the vitamin or other aspects of the nutritional content by altering the way it is grown. The amount of ascorbic acid, for example, cannot be made to equal that in an orange by the addition of any amount of commercial fertilizer or other chemical stimulant. A food's vitamin content is largely genetically determined.

Second, the distinction made between "natural" fertilizer (usually animal manure or green compost) and the commercially prepared variety is very misleading. Organic followers may try to convince you that the fertilizers used to grow their foods are inherently rich in the types of vitamins and minerals on which plants thrive while the commercial variety is sadly deficient, contaminated with artificial substances which would insult any self-respecting plant. But this is not true: all fertilizers, whether they are commercially processed, or derived from living organisms, must be broken down to their inorganic components, such as phosphorus, potassium, and nitrogen, before they can be absorbed. When the fertilizer is transformed into the nutrients that will promote growth, even the most discerning of plants cannot identify their origin. The only aspects that are usually rich about organic food products are the farmer and the merchant who produce and sell them.

So the source of a fertilizer—"natural" or "artificial"—is irrelevant. Chemicals are chemicals no matter what their origin. But furthermore, not only is natural fertilizer *not superior* in quality, it is often lacking in the types of nutrients necessary for ideal farming conditions. Processed fertilizers, on the other hand, can be consistent in quality and can correct for any deficiencies that may exist in the soil.

Additionally, there is evidence that going natural can bring with it some health risks that are easily avoided using conventional farming and food-processing methods. For instance, the use of animal or human manure in natural, organic fertilizers may lead to some real health problems. Dr. Jean Mayer, a professor of nutrition at the Harvard School of Public Health, points out:

> Biologically speaking, they [organic foods] tend to become the most contaminated of all. Organic fertilizers of animal or human origin are obviously the most likely to contain gastrointestinal parasites.

Dr. Mayer's statement was reinforced by reports from South Korea and Holland, where human sewage was being used to fertilize crops. In these countries significant portions of the populations (95 percent in South Korea) suffered from roundworm, hookworm, and other parasites. These diseases resulted from improper use of organically derived fertilizer and an incomplete chemical breakdown of the human and animal wastes. With commercially processed fertilizer, this risk is not present.

Third, the obsession with avoiding "unnatural" pesticides overlooks the very basic fact of life that insects compete with people for food. And on many occasions they have been known to win, since after all, there are significantly more insects than there are people. Their reproductive style would make any representative from Zero Population Growth cringe with dismay: it has been estimated that, if all of the offspring lived and reproduced normally, one pair of creatures from the insect world could produce 191 quadrillion (hungry) descendants in one summer season. The use of pesticides makes farming efficient and keeps consumer prices down.

The fear of some type of poisonous transfer from the spray to human food has been blown far out of proportion: agricultural scientists who use pesticides are strictly regulated by government codes on the amount of residue that may be present in their products. The levels of residue that are permitted are so small that their presence would have to be multiplied by a factor of 100 or more to reach a level that might possibly be harmful to humans. There has never been any authenticated case of illness resulting from a pesticide residue on foods purchased at retail stores.

But these are facts. And no legitimate food faddist is ever deterred by facts.

MYTH #2: Natural is by definition "good"; "artificial" is inherently suspect.

REALITY

The dichotomy of natural vs. artificial is so fallaciously fascinating that a whole chapter could be written about it. As a matter of fact, one has been (see Chapter Four, "Beware! It's Natural").

But with specific regard to organically grown foods, two 100% natural myths should be revealed here.

First, just because a product is "processed"—for instance, canned, jarred, or frozen—does not mean that it has been stripped of all its "natural goodness." It is true that processing can alter the nutritional status of a product; commercially made white bread, for instance, does lose some of its wheat germ and vitamin B. But what naturopaths tend to overlook is that enrichment is added. The final product, in terms of nutritional qualities, is often superior to the homemade, unbleached variety.

MYTH #3: Some foods have "magical" preventive or curative properties.

REALITY

The claims of "wonder foods" have no scientific basis, although the appellation "health" food certainly suggests otherwise.

Again, this is nothing new. Cato the Elder ate cabbages all day. Health faddists now concentrate on honey and wheat germ. Dr. Jarvis in 1958 recommended apple cider "to start the fat-burning process." And the same notion is still with us today—but instead of apple cider, modern-day faddists recommend a grapefruit. Perhaps ten years from now the premeal fat-burning aperitif will be a boiled okra—or a dry martini.

Probably the leading miracle food of the day is honey. One obviously enthusiastic health food store owner said, "I recommend it [honey] for ulcers, cancers, and mostly for healthy people as a preventive of both." Others proclaim it as the "sure cure" for sore throats, coughs, and colds.

Promoters of a honey-related product, Royal Jel, described as a *"pièce de résistance* of the queen bee," claimed that their product caused chickens to lay twice as many eggs and helped older chickens to start laying again. Postmenopausal women after eating Royal Jel were said to regain their fertility. "Imagine the joy of the 50-year-old women!" cried the advertisements for Royal Jel. Imagine indeed!

Wheat germ has received its share of undocumented credits. One Michigan company claimed the kernel of wheat was effective in treating neuritis, arthritis, and constipation. Given that there was not a kernel of truth in this statement, the FDA quickly moved against that group. And the FDA is constantly seeking court orders either to halt false advertising or to recall various "health" foods that offer cures for cancer, ulcers, varicose veins, and tooth decay.

MYTH #4: Vitamin E cures and prevents all diseases.

REALITY

Vitamin E has no special preventive or curative powers and routine supplemental use of this vitamin is not recommended.

One Food and Drug Administration representative refers to vitamin E as "the vitamin of the year." Books have been written extolling its virtues. If you believed everything you read about it, you'd be convinced that this capsulized miracle worker promotes physical endurance and sexual potency, improves sperm quality, prevents miscarriage, protects lungs from air pollution, and both prevents and cures heart attacks, muscular dystrophy, ulcers, cirrhosis of the liver, and cancer.

Actually all those claims about vitamin E do have something very significant in common: they are all undocumented.

Faddists have frequently resorted to emotionalism to emphasize their beliefs about the importance of vitamin E supplements. Cathryn Elwood, in her book *Feel Like a Million* (1957), refers to "expert" endorsement of high daily vitamin E doses, and points to the death of a thirty-eight-year-old father and the subsequent guilt feelings of his wife:

> She killed him, but she loved him. . . . If only I could tell them how vitamin E prevents muscles from becoming riddled with holes and torn with lesions that fill with water. This was the cause of their daddy's enlarged heart.

One of the first claims about the relationship of vitamin E and heart disease was made in 1946 when Drs. Evan Shute (an obstetrician-gynecologist), Wilfred Shute (a heart specialist) and Albert Vogelsang claimed that large doses of concentrated vitamin E were beneficial in cases of arteriosclerotic, hypertensive, rheumatic and coronary heart disease. Since then, no published or verbal report has indicated that vitamin E has value in the treatment of heart disease. The American Heart Association, the American Medical Association, the Food and Drug Administration and other well-respected medical groups have repeatedly reported that vitamin E has no value in heart disease or in any other condition with the exception of vitamin E deficiency (see below). But Dr. Shute and others still write of the glories of this "wonder" vitamin and, despite the overwhelming evidence to the contrary, unsuspecting readers still believe the material presented in such books as *The Heart and Vitamin E and Related Matters* (Evan Shute, 1969, published by "The Shute Foundation for Medical Research").

Vitamin E is a fat-soluble chemical in the alcohol family. Its chief function is that of an antioxidant—it inhibits the combination of a substance with oxygen, and thus acts as a preservative. Specifically, vitamin E inhibits the oxidation of ascorbic acid (vitamin C), vitamin A, and certain fats, enabling these essential nutrients to perform their specific functions in the body.

Vitamin E is necessary for health but supplements won't make you any healthier—that is, unless you happen to be a white rat. It seems that the claims of vitamin E's wonders are directly derived from laboratory animal experiments where vitamin E deficiencies were induced. Diseases did develop among these animals. One particularly noteworthy problem was that the sexual potency of male rats appeared to wane. Health enthusiasts immediately drew analogies to humans. (Deficiencies of vitamin E in rats also results in scaly tails, but no analogy has yet been drawn here.)

Their arguments, however, were faulty in two ways: first, scientists have established that vitamin E is so widely distributed in the foods we eat (vegetable oils, whole grains, leafy vegetables) that it is almost impossible to develop a deficiency. Second, there is no validity to the notion that if a deficiency of something causes a problem, an excess of the same substance will enhance the same potential. The Recommended Daily Allowance for vitamin E is 15 IU* for adult men and women and 30 IU for pregnant and nursing women. But the "supplement" products on the market contain 100, 200, and even 400 IU. To the best of our knowledge, ill health in humans in the United States of America has never been associated with a lack of vitamin E, nor has it been improved by giving extra amounts of vitamin E. Undoubtedly, however, you will continue to see this product labeled "Vitamin E: The Miracle Worker."

The increased popularity of this vitamin (it's been estimated that vitamin E sales doubled between 1970 and 1972) and our incomplete knowledge of the side effects of overdoses are a source of concern to public health officials.

The National Academy of Sciences has deplored the widespread fascination with vitamin E, pointing out that there are really only two types of individuals who benefit from supplemental therapy: premature babies who may have received inadequate amounts of vitamin E before

* International Unit, roughly equivalent to a milligram.

birth, and persons with intestinal disorders where fats are poorly absorbed. When used in excess, vitamin E has caused headache, nausea, fatigue, giddiness, and blurred vision in some takers. But not enough is known about other long-term effects it may have. We still have a great deal to learn about vitamin E, its role in health, and its potential toxicity at high levels. As summarized by Dr. Philip L. White, director of the AMA's Department of Food and Nutrition, "Vitamin E is still a vitamin looking for a disease."

And Dr. Lavell M. Henderson, professor of biochemistry at the University of Minnesota warns that:

> Self-medication with vitamin E in the hope that a more or less serious condition will be alleviated may indeed be hazardous, especially when appropriate diagnosis and treatment may thereby be delayed or avoided.

They and other health scientists warn that those who are supplementing their normal diet with vitamin E are taking part in a massive, uncontrolled experiment.

MYTH #5: Extra doses of vitamin A can contribute to strength and vitality, and you can't have enough of them.

REALITY

A well-balanced daily diet supplies the necessary amount of vitamin A and scientific data indicate that excessive intake of this vitamin can be dangerous (see Chapter Four for details).

MYTH #6: Vitamin C ensures good health and high doses of it are recommended.

72

REALITY

A well-balanced daily diet supplies the necessary amount of vitamin C and new scientific data indicate that excessive amounts may be dangerous.

Vitamin C is another vitamin that has set off a controversy. It has been known for many years that vitamin C plays an important part in health and that the body doesn't need a great deal of it to be healthy—generally a glass of orange or grapefruit juice, a serving of tomatoes (surprise!) or some other types of vegetables will provide the necessary daily requirement.

The real controversy about vitamin C occurred a few years ago when Nobel Prize Winner Linus C. Pauling presented his theory that massive doses of vitamin C would both prevent and cure colds. There was a great flurry of interest in his ideas—and a great deal of professional criticism of his claims. The *American Journal of Public Health Association* criticized him for making such an announcement in a popular book, as opposed to a professional medical journal, stating:

> Professor Pauling would have been more prudent and would have rendered a greater public service had he presented his ideas to the scientific world for evaluation before recommending them to the public as a basis of action.

Both before and after the publication of Professor Pauling's book, *Vitamin C and the Common Cold*, dozens of attempts were made either to confirm or to deny the health-promoting effects of supplemental vitamin C. One study at the University of Minnesota showed that a sample of people receiving high doses of vitamin

C experienced 1.9 colds per year—as opposed to 5.5 the previous year. That sounded pretty good until it was pointed out that the control group—those who received no vitamin C supplements—had an average of 2.2 colds versus 5.9 in the previous year. Somehow, a difference of 65.5 percent fewer colds with massive amounts of vitamin C as compared with 62.7 percent fewer without it, is not all that impressive.

More recently (March, 1975), two reports published in the *Journal of the American Medical Association* concluded that vitamin C shows little merit in treating the common cold. One report covered a nine month experience at the National Institutes of Health in which healthy volunteers took pills daily. Half of the subjects took three grams of vitamin C daily while the other took inert pills. The dose was doubled whenever the volunteer thought a cold infection was getting started. Three grams of the vitamin is roughly the amount one would get from eating 100 oranges. The results showed that the effects of the vitamin on the number of colds "seem to be nil."

The other report was a review of many previous studies from 1938 to present. In this report the authors concluded that the use of the vitamin for the common cold "cannot be advocated on the basis of the evidence currently available."

Not only does it appear that daily doses of vitamin C have no beneficial effects with regards to colds, but the self-medicated patient may be assuming some risks. Professor Pauling recommends up to 20–100 times the Recommended Daily Allowance (RDA) for preventing colds —and up to 200–400 times the RDA for curing them. But what type of side effects might occur with such massive doses?

It has been established that some people on these high-dose regimens develop diarrhea. Recent studies, however, have suggested that massive doses of vitamin C may also increase the chance of developing kidney or bladder stones, may in some way affect bone metabolism, and could have a negative effect on the status of vitamin A.

An Australian physician has advanced the hypothesis that large amounts of vitamin C may reduce female fertility by changing the nature of the secretions around the cervix, making it difficult for sperm to enter. Furthermore, questions have been raised about the effects excessive doses of this vitamin can have on the fetus. There is some evidence that vitamin C overdose during pregnancy can lead to increased requirements for vitamin C in the newborn, increasing his chances of developing scurvy despite a normal diet.

Most physicians today do not recommend vitamin C overdosing. Not only do they feel it is a waste of money, but again, the side effects are at this point unknown.

Professor Pauling is one of only a few individuals twice honored with the Nobel Prize (chemistry and peace). But he is not a physician and not a nutritionist.

MYTH #7: Lecithin is nature's protection against heart attacks.

REALITY

There is no evidence that lecithin protects against heart attacks or any other disease.

Lecithin is a natural emulsifying agent. It has been used for many years as a food additive in mayonnaise, salad dressings, and chocolate to maintain the stability of the oil ingredients.

Recently lecithin has been glorified as the key to preventing heart attacks. Some food faddists contend that lecithin can cure arthritis, high blood pressure, and gall bladder problems, while also improving brain power. Others recommend it in conjunction with cider vinegar, kelp, and vitamin B complex as a weight-reduction method. But the primary asset of lecithin is its alleged capacity to ward off coronary heart disease.

The theory is that lecithin breaks up and disperses cholesterol in the blood so it cannot become attached to the artery walls. But, as with so many other health claims, there is no medical evidence to back up this belief.

The body manufactures its own lecithin and there is no reason to believe that any additional quantities of this natural emulsifier are necessary or helpful. Dr. Eleanor Williams, associate professor of nutrition at the State University College at Buffalo, has criticized the current trend toward lecithin supplements, pointing out that the exact mechanism whereby cholesterol becomes deposited in the arteries is unknown. It is, she feels, totally misleading to assert that lecithin will prevent such deposits.

Heart specialists have confirmed that lecithin has never been shown to prevent arteriosclerosis and these doctors have expressed concern that those who are buying lecithin in capsules, chewable wafers, tablets, and oil solutions are deluding themselves—continuing to smoke, omitting exercise and proper diet—that is, not giving attention to the variables that *have* been linked with higher probabilities of heart disease.

There is no evidence at this point that excessive doses of lecithin actually do any harm—except perhaps to your food budget.

THE FDA AND THE VITAMIN CONTROVERSY

Each year Americans spend more than 300 million dollars on vitamins—and many of these "believers" operate under the assumption that you can't have too much of a good vitamin. Given the toxic effects of certain vitamin overdoses, the FDA is cracking down on some types of vitamin sales, a move that has made Counsel Hutt very unpopular with the health food industry and earned him the title of "number one bad man."

First, the FDA has taken the position that an industry cannot combine useless ingredients with necessary ones. For instance, it is now illegal to combine an essential nutrient, such as vitamin C, with rutin, a tasteless greenish-yellow powder from buckwheat and other sources—and then sell it at a significantly higher price than vitamin C alone would bring. To do so would be fraud since the consumer would think that if vitamin C is good alone, in combination with rutin it must be better. When the FDA made this decision, it received over 100,000 letters from consumers of rutin–vitamin C capsules who were irate over the FDA's attempted "interference with their health." Mr. Hutt's response was: "I don't care if people eat rutin . . . they can have it for breakfast, lunch and dinner, they can have it in place of coffee, in their snacks. . . ." His only concern was that it was being misrepresented in its expensive combination with vitamin C. Rutin, alfalfa, and bioflavonoids and other substances that had previously been mixed with essential nutrients are still available by themselves.

Second, and more serious, is the FDA's concern about excessive vitamin intake. The prevailing notion regarding vitamins appears to be "if some is good, more is better." Unfortunately for health food enthusiasts, however, it doesn't always work that way. While excess intake of some vitamins (the water soluble ones, B vitamins and C) is naturally excreted by the body (it is said that the richest source of vitamins is the New York City sewage system), others (the fat solubles, A and D) can cause serious damage.

Connecticut Medicine carried an account of a thirty-one-year-old woman who was in a very depressed condition and, on the counsel of her mother, went to the local health food store for help. She was advised to take large quantities of "natural vitamins and health food supplements," and further told that "shaking the head side to side 25 times, forward and backward 20 times and diagonally 15 times" would make her feel much better. Her daily intake of medication is listed below:

DAILY MEDICATIONS TAKEN BY PATIENT

# Tabs	Medication and Composition/Tab
6	*Acidophilus & Pectin*—Lactobacillus Acidophilus Extract, Pectin (300 mg)
6	*Nuplex*—Vit. A (25,000 units), Vit. D (400 units), Vit. B_1 (10 mg), Vit. B_2 (10 mg), Vit. B_6 (5 mg), Nicotinamide (100 mg), Vit. C (200 mg), B_{12} (5 mg), Vit. E (5 units), Calcium (125 mg), Iodine (15 mg), Iron (15 mg), Copper (1 mg), Magnesium (6 mg), Manganese (1 mg), Zinc (1.5 mg)
4	*Iron With Supplement*—Vit. B_{12} (25 mg), Iron (115 mg), Copper (0.5 mg), Manganese (2 mg), Vit. C (30 mg), Lemon Flavonoid (30 mg)
6	*Vitamin E*—100 units
6	*Manganese*—2 mg
6	*Vitamin C*—Vitamin C (500 mg)
6	*Vitamin A and D*—Vit. A (15,000 units), Vit. D (2,000 units)

6   *Calcium Pantothenate*—100 mg
9   *Enzymall*—Assorted Pancreatic Enzymes
3   *Formula* 104 *Multiple Minerals*—Ferrous Sulfate (65 mg), Calcium Phosphate (30 mg), Potassium Iodide (0.15 mg), Magnesium Oxide (12 mg), Sulfur (15 mg), Zinc Sulfate (10 mg), Copper Sulfate (5 mg), Potassium Chloride (5 mg), Manganese Sulfate (5 mg)
6   *Bone Meal*—Calcium Phosphate (750 mg—340 mg as elemental phosphorus)
6   *Dolomite*—Calcium (130 mg), Magnesium (78 mg)
6   *Glutamic Acid HCL*—Equivalent in Hydrochloric Acid content to 10 drops of Dilute HCL U.S.P.
6   *Vitamin B Complex Formula* #49—Vit. B_1 (5 mg), Vit. B_2 (5 mg), Vit. B_6 (5 mg), Folic Acid (0.1 mg), Pantothenic Acid (20 mg), Para Amino Benzoic Acid (30 mg), Niacin (50 mg), Inositol (1,000 mg), Choline (1,000 mg), Vitamin B_{12} (15 mg), Ciotin (25 mg)
6   *Potassium Chloride*—75 mg
6   *Folic Acid*—10 mg
8   *Chlortrimeton*—4 mg
4   *Prolixin*—5 mg
3   *Thyroid Extract*—¼ gr
6   *Unsaturated Fatty Acid*
6   *Safflower & Linseed Oil*
6   *Wheat Germ Oil*

This unfortunate lady soon showed up at a hospital emergency room complaining that she was losing all her hair. She was taken off the vitamin and supplement diet and her hair soon returned. Presumably she decided it was better to be a little depressed than totally bald.

The loss of hair is a relatively minor side effect of vitamin overdose compared with irreparable damage to internal organs and the type of unnecessary brain surgery which has been known to follow complaints related to vitamin A overdose (see Chapter Four), and the calcification of soft tissues and bone deformity that can follow excess diet supplementation with vitamin D.

Currently, dosages of vitamin A and D that exceed certain levels may be sold only by prescription. The RDA for vitamin A, for instance, ranges from 2,500 IU for young children to 5,000 IU for adult men and women. The recent FDA ruling defines as prescription products dosages exceeding 10,000 IU. Some medical scientists feel that even that level may be too high. But the health food promoters are still annoyed, enough so to initiate lawsuits against the FDA. They interpret the government move as an infringement on their right to health. And the FDA proposal to make *all* vitamins which exceed 150% of the RDA over-the-counter drugs is being met with great resistance. As of this writing, it appears that the FDA will drop plans for such a classification, maintaining only the restrictions on excessive doses of vitamins A and D.

Natural Is Good?

The current fascination with "naturalness," including unpasteurized milk, uncooked eggs, beef, and pork, and supplements of "natural origin," is a real source of concern. It took us hundreds of years to understand how to control the elements that contribute to a healthy balanced diet and to learn how to control typhoid, salmonella, tapeworms, and trichinosis. It is puzzling why the back-to-nature movement insists on including the type of nostalgic reversions which carry with them the risks of lethal diseases.

Because the current flight from food additives and the

proliferation of health food stores are based on the premise that "natural is better," it is worth taking a look at the not so well known facts about Mother Nature's own products. Read about the "natural chemical feast," and weep.

FOUR

Beware! It's Natural

"When you next go tripping through the
Garden of Eden, be sure to take a qualified toxicologist with you."

AN ANONYMOUS, ALTHOUGH
VERY WISE, OBSERVER OF
THE RETURN-TO-NATURE
MOVEMENT

IF YOU CAN'T PRONOUNCE IT, IT MUST BE HARMFUL

THE next time a natural-food faddist stops in for a
visit, be sure to offer him a hot, steamy solution which
contains, among other things, caffeine, tannin, essential
oils, butyl, isoamyl, phenyl ethyl, hexyl and benzyl alco-
hols, and geraniol. Undoubtedly, he will shiver in abhorrence, and politely turn you down. He would probably
be embarrassed to learn that he was declining a simple
"natural" cup of tea.

Underlying the natural foods movement is the inaccurate idea that "artificial" foods are made up of "chemicals" while natural foods are "chemical-free." After all,

who wants to eat some treacherous-sounding laboratory creations with a name you can't even pronounce? Butylated Hydroxyanisole (BHA), Butylated Hydroxytoluene (BHT). Sodium Bisulfite. Xanthan Gum. It's enough to make you get a one-way ticket and take the next available conveyance to Healthfoodland.

Just keep it natural, say health faddists. None of that chemical stuff for me. All I want is an organically prepared, naturally cooked breakfast—say some chilled melon, scrambled eggs, and a nice, freshly brewed cup of coffee.

From the point of view of the chemist, the health faddist had just ordered the breakfast described on page 84.

The assumption that natural food is chemical-free is absurd. Every living thing is made of chemicals. Even your best friend is 65 percent oxygen, 18 percent carbon, 10 percent hydrogen, 3 percent nitrogen, 1½ percent calcium, 1 percent phosphorus, and 1½ percent gold, silver, and other elements. The human body is also composed of a small amount of arsenic, 15–20 mg. to be exact. But the unrealistic dichotomy between natural and artificial is perpetuated.

Gerald Gold of the *New York Times,* in reviewing Fleischmann's Egg Beaters (cholesterol-free egg substitute) and other new food substitutes, noted that "you would have to be a chemist to tell what you are really eating, even though all the products list their ingredients, for some of them are mostly chemicals." He then pointed out the Egg Beaters contained "egg white, corn oil, nonfat dry milk, emulsifiers (vegetable lecithin, mono and diglycerides and prophylene glycol monostearate), cellulose and xanthan gums, tri-sodium and triethyl citrate, artificial flavor, aluminum sulfate, iron phosphate, artificial color, thiamin, riboflavin and vitamin D."

A CHEMIST'S VIEW
OF BREAKFAST*

Chilled Melon
Starches
Sugars
Cellulose
Pectin
Malic Acid
Citric Acid
Succinic Acid
Anisyl Propionate
Amyl Acetate
Ascorbic Acid
Vitamin A
Riboflavin
Thiamine
Phosphates

Scrambled Eggs
Ovalbumin
Conalbumin
Ovomucoid
Mucin
Globulins
Amino Acids
Lipovitellin
Livetin
Cholesterol
Lecithin
Lipids (Fats)
Fatty Acids
Butyric Acid
Acetic Acid
Sodium Chloride
Lutein
Zeaxanthine
Vitamin A
Phosphates

Coffee

Caffeine
Essential Oils
Methanol
Acetaldehyde
Methyl Formate
Ethanol
Dimethyl Sulfide
Propionaldehyde

Acetone
Methyl Acetate
Furan
Diacetyl
Butanol
Methylfuran
Isoprene
Methylbutanol

* Note: 1) The above chemicals are chemicals found *naturally* in food. 2) The chemical listings are not complete. 3) Used with permission of the Manufacturing Chemists Association, Washington, D.C.

"Here you've just had a nice low-cholesterol, low-cal breakfast of egg white, corn oil, skim milk, lecithin, mono- and diglycerides, propylene glycol monostearate, cellulose and xantha gums; trisodium and triethyl citrate, fortified with thiamin, riboflavin, and vitamin D; decaffeinated coffee with nutritive lactose and soluble saccharin .. and you're still not happy?"

Someone with even a mild degree of chemical-phobia, upon reading this array of polysyllabic words, would probably elect to keep his cholesterol high rather than risk the unknown terrors of becoming a walking test tube. If, however, the same person took the time to find out what was in an egg, he might begin to wonder if the chemicals in this natural product were capable of scrambling his genes.

One of the most unrealistic and misleading of all applications of the artificial vs. natural dichotomy has been in the field of vitamins. Health food promoters have attempted to convince their customers that "natural" vitamins are better. In reality, of course, vitamins are specific chemical compounds. Your body can't tell the difference between a vitamin that was synthesized by a chemist or derived from a natural source. Again, only the merchant can tell the difference—by comparing the margin of his profits.

THE POISONS IN YOUR HEALTH FOODS

Those people who seek shelter in natural foods assume that, by definition, they are safe. But in making that assumption, they are overlooking the results of many years of scientific research.

If a food faddist really goes all out and assumes that anything that is present in nature is eligible, in unlimited quantities, for his health food shopping list, he's in big trouble. Consider some examples.

Vitamin A

Vitamin A is essential to good health. It plays a vital part in overall vision, children's growth and bone development, the health of the skin and mucous membranes. Everyone knows that vitamin A helps you avoid the problems associated with seeing in the dark (a particularly important asset in the time of an energy crisis!). Vitamin A is found (among other places) in eggs, butter, cream, fortified milk, meats, and liver; deep yellow and leafy green vegetables contain carotenes, substances that the body can convert to vitamin A.

Most foods that occur in nature do not have enough vitamin A to bring about poisoning or vitamin A intoxication. But in 1856 an Arctic explorer documented an exception to that rule. He and his companion feasted one night on a healthy serving of polar bear liver, and shortly after, recorded symptoms of vertigo, violent frontal headache and nausea, drowsiness and irritability. A century later, diners enjoying shark liver reported the same effects. Both groups had eaten food that contained an extraordinary amount of vitamin A and they were experiencing its toxic effects.

Today, while few people include polar bear or fish livers in their diet, many are taking very high supplementary doses of vitamin A. And physicians are reporting side effects. Of particular concern is the observation that those suffering from vitamin A intoxication report the same type of intracranial pressure and violent headaches that are associated with brain tumors. More than once a serious and unnecessary operation has been performed. In addition, there have been deaths from cirrhosis of the liver among food faddists who have used inordinate amounts of carrot juice each day to wash down their high-dose vitamin A capsules, and incidences of serious dry skin disorders, hair loss, and abdominal pain have been associated with abuse of vitamin A.

Those lovers of carrot juice who may drink a quart a day of it for weeks on end not only run a risk of having an excessive amount of vitamin A but also of producing a condition known in humans as carotenemia, the presence of the coloring matter of the carrots in the blood, which causes the palms of the hands and soles of the feet to become yellow. But there's more.

When administered in high doses to pregnant animals, vitamin A has been shown to cause birth defects, including cleft palate and brain deformities. Similarly vitamin A has been shown to bring about breast cancer in animals. The type of experiment that induces animal cancer with vitamin A overdoses is almost identical to the type that has led to the banning of many so-called "artificial" foods (cyclamates, for example).

87

Using three populations of mice (one overall control group that did not receive vitamin A supplements, one experimental group given a high vitamin A content in their diet, and another control group given a regular diet, but in supervised amounts to maintain their weight at the level of the vitamin A-saturated mice), it was reported that "within the limitations of the experimental system, data indicate that a significantly higher percentage of animals in Group II (those receiving excess vitamin A) developed pulmonary metastases. . . ."

But no one seems to be concerned about moderate use of vitamin A. And indeed health food fans are very agitated that the FDA has now forbidden the over-the-counter sale of excessive amounts of this vitamin. At the time of this writing, these protesters are conducting a huge letter campaign to Congress with the intent of nullifying FDA action in this area.

Aflatoxins

In 1960, thousands of turkeys died in England and elsewhere of what was temporarily called "turkey X disease." Soon after, it was learned that natural aflatoxin molds (*Aspergillus flavus*) were the causative agent.

Aflatoxin-producing strains are widely dispersed in air and soil, and have the capacity to grow on a variety of substances, including peanuts, rice, corn, soybeans, whole oats, and wheat (particularly shredded wheat). The cancer-causing effects of aflatoxins have been demonstrated in rabbits, guinea pigs, dogs, cattle, ducks, rhesus monkeys, and mice. For instance, at a dietary level of .030 parts per million (ppm), aflatoxin contained in toxic groundnut meal produced liver tumors in ducks within fourteen months.

There is now some serious consideration being given

to the possibility that aflatoxins in moldy peanut products may play a role in the development of human liver cancer. An epidemiologic study in Swaziland (Africa), where liver cancer is the most common of all malignancies among men, linked the prevalence of this disease with the tendency to eat infested—but very natural—groundnuts.

A 1969 study in the Philippines noted that a sample of locally manufactured natural peanut butter was highly contaminated with aflatoxins. And it is hardly comforting to learn that Dr. Wodicka has admitted that "trivial amounts" of natural aflatoxins can be found even in those U.S. peanut products that do get to market. A recent FDA nationwide study of many brands of peanut butter found aflatoxins in 25 percent of the peanut butter tested but in amounts too small to cause ill health in humans. Generally, commercial peanut products sold in this country have aflatoxin traces below 5 parts per billion, but these traces can and often do remain despite cooking, processing, or home-roasting peanuts.

The FDA does carefully inspect all peanut products to be used for human and animal consumption and has, on occasion, rejected some batches that looked suspicious. But some traces still appear. The puzzling factor is the relative lack of concern over "natural" mold and the havoc it can cause with health (minor though the chances are that it would do so) and the preoccupation with even the hint of a problem associated with an "artificial" product.

"Chickpeas" (*Lathyrus sativus*)

Now what could be more natural and better than a bowl of freshly picked chickpeas. But wait. Overindulgence in chickpeas, or, specifically, seeds of the *Lathyrus sativus,* has been linked to the development of lathyrism, a disease of the central nervous system characterized by muscular weakness, which can cripple its victim for life.

In Matthew 13:25, there is a reference that reads: "But while man slept, his enemy came and sowed tares among the wheat." Some interpreters feel that this is a reference to chickpeas and their devastating effect on the lower limbs.

In several villages in central India, entire male populations are affected by this disease. Lathyrism is as widespread as it is because of the fact that the chickpea, a hardy vegetable, can survive droughts and other forms of poor weather, and is available when every other source of food fails.

Because menus in the United States and Western Europe do not generally feature *Lathyrus sativus,* lathyrism is not a problem in these areas. The "chickpeas" we come in contact with are of a very different variety than are those, for instance, in parts of India and thus they do not pose a risk to health.

Ergot

Certain types of natural rye and wheat have a special susceptibility to the development of a parasitic fungus known as ergot.

The pharmacologic effect of ergot has been known for hundreds of years: it causes muscles and blood vessels to contract (and thus has been used to treat migraine headaches and induce abortion) and in large doses may constrict capillaries to the point of making a person's hands and knees numb, producing gangrene and violent death.

The active components of ergot are a number of alkaloids built around the basic nucleus of lysergic acid

(LSD is lysergic acid diethylamide). There are episodes in history where the LSD-like effects of ergot ("St. Anthony's fire") have caused death and serious injury: in the year 944, it is said that over 40,000 people died from the consequences of eating ergot mold. In 1951, there was an outbreak of "bread poisoning" in Pont Saint-Esprit, near Avignon, in southern France. After eating the natural products of a local bakery, over two-hundred people became ill, four died, and twenty were classified as "temporarily insane."

Some of the victims tried to fly off buildings. Others screamed in terror that they were being surrounded by fire and attacked by prehistoric beasts. The effects of the ergot made colorful material for newspaper writers the next morning, but for a couple of hundred people who were directly involved, it was one bad, and unanticipated, trip.*

In addition to its potential for inducing a St. Anthony's fire effect, natural ergot has also been linked with tumor formation in certain laboratory animals.

Because of these potential problems, rye and wheat products are now routinely examined in the United States, and ergotism poisoning appears not to be a serious health threat. Major bread industries know about it, and they know how to guard against it. But when Dr. Virgil Wodicka was asked about the possibility of a resurfacing of the ergot problem related to the back-to-nature movement he stated: "Well, I'd say that there'd be that chance, because the people who push this natural bit ordinarily are the marginal operators who don't know

---

* There is some controversy as to whether or not ergot alone was responsible for the Pont Saint-Esprit incident. There is one hypothesis that mercury poisoning was responsible, or at least involved.

very much about what they are doing, and they're therefore not careful."

Sassafras and Apricot Kernels

Safrole, a component of the natural sassafras plant and
oil of sassafras, has been shown to be a cancer-causing
agent. It was used until 1960 as a flavor for root beer.
After all, it was a natural substance, so how could it be
suspect? But experiments confirming that it caused liver
cancer in some animals led to its demise as a natural
food additive.

But now sassafras is back as the basis for a "health
food" tea. When the health food store proliferation occurred in the early 1970's, the FDA officials were
alarmed to find that fresh sassafras roots, newly pulled
from the ground, were popular consumer items. They
were quickly seized from the market, and sassafras tea
products are now allowed only under the condition that
the active ingredient, safrole, has been removed.

Another health food store's "purely natural" product
that has posed this same type of problem is aprikern—
quite literally, the kernel of the apricot.

Unfortunately for those who are indiscriminate about
eating natural, apricot kernels contain hydrogen cyanide.
Recently a three-year-old girl suffered from cyanide poisoning after eating approximately fifteen of these natural
"delights."

Spinach

Popeye would be very distraught to hear the things
that are now being said about spinach.

Our major intake of nitrates in foodstuffs comes primarily from vegetables. A person is likely to consume
more nitrates from his vegetable intake than from cured

meat products he eats (see Chapter Seven). Nitrates are a natural plant constituent, but they occur in extraordinary amounts in spinach, beets, radishes, eggplant, celery, lettuce, collards and turnip greens. The content of some samples has been shown to be more than 3,000 parts per million. Nitrates have the capacity to be converted in various body tissues to nitrites—and nitrites, under some circumstances, have been shown to be toxic.

Usually the conversion of nitrate from vegetables to nitrite either does not occur, or if it does, presents no known problem. But there is an important exception to this rule: when spinach, whether processed or "natural," is stored under conditions that permit the growth of microorganisms, nitrate may be changed to nitrite.

A number of cases of infant "spinach sickness" have been reported following the ingestion of fresh spinach that was left at room temperature for some time after cooking. Experts in the area of food chemistry now recommend that:

Home-prepared spinach should never be stored for subsequent feeding . . . [and] in view of the apparent sensitivity of young infants, prudence would dictate that foods such as spinach and beets, containing high levels of nitrate, should not be introduced into the diet of children below three months of age.

So, remember when your mother said "eat-your-spinach-it's-good-for-you"? Evidently she didn't know about spinach poisoning (or she was using fresh spinach) or about some of the hypotheses that the oxalic acid in spinach and rhubarb may contribute to the formation of common types of kidney stones. (Poor Popeye. Passing stones.) Spinach, of course, is an excellent, nourishing

93

food when it is eaten in moderation while it's fresh—but perhaps Clarence Darrow suspected its potential problems when he wrote: "I don't like spinach and I'm glad I don't because if I liked it I'd eat it and I'd just hate it."

A NATURAL GOURMET REPAST

Imagine being invited to an elegant dinner party given by a gourmet naturopath. The menu might read something like this:

ASSORTMENT OF NUTS

FISH COURSE
Porcupine Fish
(fresh from the sea)

MAIN COURSE
Roast Stuffed Quail
Yellow Rice
Sweet Potatoes

SALAD
Collard Greens and
Mushrooms tossed with
Young Fresh Shoots of
Bracken Fern

APPLE CIDER
(organic, of course)

COFFEE

Look tempting? Better read before you eat. Check over those "natural" nuts very carefully. In addition to being wary of aflatoxin growths on the peanuts if they are not properly preserved, make sure your naturopathic host isn't serving you any cycad nuts, the product of palmlike trees that still inhabit tropical and subtropical lands.

Because the cycad nut is able to survive droughts and hurricanes, it has traditionally supplied a readily available source of food in some parts of the tropics, and has also provided a basis for various medicinal products. When rats are fed cycad nut meal, they develop primary tumors in their liver and kidney.

94

"Porcupine" fish, also known as fugu fish and balloon fish, is a still-famous Japanese delight, which derived its various names from the fish's capacity to inflate its prickly hide when it is unduly annoyed. Throughout history, hundreds if not thousands of people have been poisoned to death by this delicacy. But people go on eating it—it's natural.

Ingestion of just a small bit of toxic material from this and other types of fish can bring about numbness of the lips, tongue, and fingertips in a matter of minutes. The potent poison is concentrated in the liver, ovaries, and to some extent the testes of these sea creatures. The Japanese try to have their fish and eat it too by licensing chefs who are formally trained in preparing this dish. They learn how to identify the toxic organs and discard them. That is, usually they discard them. Some Japanese men like to take the chance of eating the testes portion anyway—as a means of enhancing their virility. In 1970, forty-two fatal cases of "fugu fish" poisoning were reported in Japan.

So you'll pass on the fish dish. On to the quail, a favorite game bird in southern Europe and along the African coast. Unfortunately, however, quail is the subject of one of the oldest known cases of food poisoning.

The Bible mentions that the hungry Hebrews were saved by quail on a number of occasions, but in some instances the bird did prove to be poisonous.

> But before they had sated their craving, while the food was still in their mouths, the anger of God rose against them, and He slew the strongest of them and laid low the picked men of Israel.
>
> (Psalm 78)

Presumably, the victims were being punished for their greediness in overeating. The observation that the meat was still in their teeth suggests a rather acute form of poisoning.

In later years, Pliny the Elder, Lucretius, Galen, and Avicennus mention the dangers of eating quail. These writers seemed to feel that the lethal nature of quail was due to their tendency to eat poisonous substances, such as hemlock. (Some quail survive after eating large doses of hemlock. When the meat of quail is fed to dogs surviving such feeding tests, the results are fatal.) Whatever the origin of the toxic substances, it appears that certain quail contain coniine alkaloids which can be fatal if eaten by man. In Algeria, where a so-called green quail is eaten (i.e., quail returning in the spring from central Africa to Europe and often quite emaciated), there are reports of people developing nausea, vomiting, cold shivers, and partial, slow-spreading paralysis, all associated with this natural—but foul—fowl delight.

No solace in rice, either, if it is not fresh. Stored rice is quite susceptible to contamination by many fungi, especially *Penicillium* and *Aspergillus* species. Included in a shipment to Japan after World War II was some rice that had become contaminated and bitter by this natural process, so-called yellow rice (not to be confused with saffron rice). It was given to rats and mice and, at high diet levels, both cancerous and noncancerous tumors developed. Effects of yellow rice were subsequently confirmed in observations of humans exposed to this food.

You'd better check out those sweet potatoes too. Large scale outbreaks of sweet potato poisoning among farm animals have been noted. Cattle, especially, have been shown to develop fatal lung edema and die from apparent asphyxia. Preliminary studies have found toxins in some store-bought moldy sweet potatoes—the types of toxins that will not disappear after boiling or baking. The moral of this vegetable course? Buy fresh products.

A tossed array of garden products! Fresh or not, that could mean trouble. As mushroom lovers are well aware, there are over twenty-five highly toxic species of natural mushrooms—including the infamous "death angel" variety, which produces rapid degeneration of the liver, kidney, heart, and muscles. Some salad vegetables (particularly cabbage, cauliflower, turnips, mustard, collard greens) if eaten in excess may induce goiter in susceptible individuals, blocking the absorption of iodine and leading to a real health problem.

And then there is bracken fern (*Pteridium aquilinum*), a tender tasty sprout, with an appearance much like asparagus, which is, in all its natural glory, a powerful cancer-causing agent, even after it is cooked. This is one vegetable that is a particular source of concern to Dr. Elizabeth Weisburger, of the National Cancer Institute. Dr. Weisburger points out that the trend to "go natural" and eat whatever looks tempting in the garden can have some dangerous—and possibly deadly—consequences: for instance, cows feeding on this natural delight suffer damaged bone marrow and polyps (swollen membranes) of the bladder.

Refreshing natural apple cider. Enjoy it? If it's derived from organically grown apples, perhaps you ought to decline. Patulin, a toxic metabolite of several fungi and a substance known to be cancer causing, has recently been identified in certain types of unfermented apple juice. According to *Nutrition Reviews,* "Juice derived from apples grown on organic farms where trees have not been sprayed is likely to contain considerable quantities of fungus-rotten apple extract. This fact should be of concern to advocates of organic farming practice."

Your coffee, at least that is good to the last drop. Or is it? Caffeine, a "natural" substance found in soft drinks as well as coffee, has been shown to have deleterious effects on bacteria, fungi, and plants, and can cause abnormalities in unborn mice if administered at high doses during pregnancy. In addition, a study of patients with cancer of the lower urinary tract revealed a relationship between coffee intake and risk of bladder cancer. The authors of this study concluded that the relationship of excess coffee drinking and disease should be further investigated. (This admonition is generally sound in any field of science when a suggestion of a health hazard is raised.)

PASS THE SODIUM CHLORIDE

Food has been described as the "most complex part of man's chemical environment." We are learning more every day about the intricacies of the makeup of some of the very natural foods we tend to take for granted. What has become clear is that any food, natural or otherwise, can be dangerous if it is taken in high dosage for many years, or as the Renaissance physician Paracelsus said, *"sola dosis facit venenum,"* which means "only the dose makes the poison."

We know that simple table salt (sodium chloride), which is present in all living things, may bring about problems if it's eaten in substantial quantities. Animal studies have demonstrated that excess salt intake interferes with growth and raises blood pressure. In humans it is known that congestive heart failure can be exacerbated by high salt levels. Toxemia of pregnancy, certain types of kidney diseases, and hypertension can

98

frequently be relieved by drastic reduction in salt consumption. Physicians in Galveston, Texas, where the natural water supply has an unusually high level of salt, are very aware of these problems.

Furthermore, natural everyday sodium chloride is habit forming and many doctors would like to have less of it used in infancy and childhood so as not to form this habit. Salt overdosing in children (which has happened when salt was mistaken for sugar) causes vomiting, fever, respiratory distress, and convulsions.

At least one study has indicated that there is a high incidence of right-sided heart lesions in a group of Africans who use bananas as their major source of food. Of course this doesn't prove that bananas are at fault and certainly doesn't mean that bananas used in our diet are harmful. Bananas, like most foods eaten, contribute to good nutrition. But it does suggest again that too much of anything—even a good food—can cause problems.

In the same context, consider potatoes, usually considered as one of man's simpler foods. Actually, however, a potato represents a complex aggregate of about one hundred and fifty different chemical substances, among them solanine alkaloids, oxalic acid, arsenic, tannins, nitrate and over a hundred other items of no recognized nutritional significance to man. What is interesting here is that the toxicant solanine in the "natural" potato is the very same substance that is found in the foliage of deadly "nightshade" (solanine is a potent cholinesterase inhibitor and can interfere with the transmission of nerve impulses). Of course most of the solanine in potatoes is near the skin, and especially in the green portion which has been exposed to the sun or even artificial light and which we should not eat.

(One researcher showed a fourfold increase in solanine levels in potatoes exposed to normal illumination levels in the supermarket.) Nevertheless in consuming 119 pounds of potatoes (average) a year, we take into our system 9,700 milligrams of solanine, enough to kill a horse. But it really doesn't matter because we don't eat that much of it at one time. Even if you are very fond of potato skins you have no need to be concerned.

And while we're talking about vegetables, consider lima beans, which are consumed at the average of some 1.85 pounds per year. Lima beans contain hydrogen cyanide (which was one of the Nazis' favorite suicide potions). But then again, under normal eating conditions we attribute no health problems to lima beans.

Adding spices to your life can be harmful if you do it in excess. Nutmeg (nutmeg is the dried kernel of the seeds of the nutmeg tree) contains the hallucinogen myristicin. An overdose of nutmeg will bring about a type of euphoria, complete with hallucinations and a dreamlike feeling. Nutmeg highs, however, are not recommended in that abdominal pain, vomiting, insomnia, incoherence, depression, a severe stupor, and possibly liver damage often follow. But even without an overdosing, we eat about one-third of an ounce of natural nutmeg each year (about two teaspoons full) and thus take in about forty-four milligrams of myristicin. And we again have identified no problems here.

Estrogen, which under some circumstances is a cancer-causing agent (this is the basis of the furor over DES), is detected in eggs and in many plants—carrots, soybeans, wheat, rice, oats, barley, potatoes, apples, cherries, plums, garlic, sage leaves, parsley, and licorice root. Estrogen can even be found in wheat germ and honey! Oh, but the health food faddist would say, this is "natural estrogen," as if it were any different.

It probably doesn't sound very appetizing, but it's also true that many human foods—for instance, fruits, vegetables, cereal products, meats, dairy products—have traces of arsenic. Food of marine origin, especially oysters and mussels, are much richer in arsenic. Prawns from the coastal waters of Britain have been shown to contain up to 174 ppm of arsenic. This arsenic occurs naturally—it is not a result of environmental pollution. But people who eat these natural products of the sea without complaint are indignant when traces of arsenic in organic compounds such as arsanilic acid, used as a growth stimulant in rations of pigs and poultry, appear in meats. (This type of arsenic is rapidly excreted and does not accumulate.) Actually, the next time you begin thinking about the possibility of arsenic in your diet, consider the fact that scientists have now documented that traces of arsenic, a basic mineral, are indeed *necessary* in the human diet for good health.

Water can be dangerous—and lethal too—especially if you find yourself totally submerged in it without benefit of supplementary air. It is also reported that a grisly Oriental custom involves suicide by means of consuming great quantities of water.

As the final blow, it should be pointed out that those hallmarks of Healthfoodland, honey, wheat germ, and yogurt, aren't always the naturally delightful substances they are supposed to be. In some cases these three wonder foods can cause health problems, too.

In addition to the fact that it has traces of estrogen, the existence of poisons in certain types of honey has, throughout history, been well confirmed. The oldest description of mass honey poisoning was made by Xenophon during the expedition of Cyrus in 401 B.C. near Trebizonde in Asia Minor.

One description of intoxication by a type of mountain laurel honey, written by the attending physician, is as follows:

> Symptoms of poisoning—these have varied in severity from a mere tingling in the skin to almost death. No two persons reacted exactly alike, yet they did enough as to be recognized from a common causative agent. Shortly after eating, within a few minutes to two hours, the person felt a tingling and numbness in the extremities and lost consciousness, sometimes but momentarily and others for several hours. The pulse weakened to imperceptibility and went down to 50 or even 30. The face turned ghastly blue so characteristic of a heart attack but without anginal pain. A cold sweat appeared. These symptoms lasted from a few minutes to four or five hours. Usually there was no nausea or other gastrointestinal symptoms. . . .

There is very little chance of encountering poisonous honey in the United States today, because high screening standards are maintained and potentially poisonous substances are eliminated from the market, but it is interesting to keep in mind that even the alleged panacea from the beehive can, on occasion, be suspect.

Wheat germ is rich in protein, vitamins, and minerals (although its contribution to the diet is questionable since it is usually eaten in small quantities). But when its fat components turn rancid, it can serve as a growth medium for various types of fungi, including the potent and dangerous ergot. Wheat germ must be kept refrigerated (which somehow seems a bit ironic—a product of nature having to be stored in a product of the notorious industrial complex).

Yogurt. The symbol of the "natural generation." If you believed everything you heard about this food, you'd be convinced that it, in itself, constituted the perfect diet. Researchers at Johns Hopkins Hospital in Baltimore decided to find out what would happen if they put rats on an exclusive diet of this wonder food. They found the rodents "avidly" ate the yogurt. They grew at a normal rate, mated, conceived, and had normal litters. But *100 percent of the yogurt-eating rats developed cataracts!* In the adult rats, the cataracts appeared four to six months after they began their yogurt spree. The researchers hypothesize that the high galactose levels in commercially prepared yogurt might be responsible for their observations.

Of interest here is that *all* the rats developed cataracts. As will be seen in subsequent chapters, when additives are tested, in a similar manner, they are likely to be banned if only a small percentage of the laboratory animals manifest health problems.

WELL, WHAT CAN WE EAT?

The purpose of the above discussion about the problems of overdosing of any food—and the presence of toxic materials in many natural foods—was not to cause alarm about the food we eat, but to indicate that there is an important distinction to be made between the words *toxicity* and *hazard*. The *toxicity* of a substance is its intrinsic capacity to produce injury when tested by itself. For instance, arsenic, lead, mercury, fluoride, and other substances have high intrinsic toxicities. A *hazard*, on the other hand, is the capacity of a substance to produce injury under the circumstances of exposure. For instance, there is no known problem about the presence of the toxic substance solanine in potatoes or hydrogen cyanide in United States* lima beans. We consume a multitude of toxic substances every day, most of which come from "natural" foods. We don't, however, get ill. But why don't we?

There appear to be at least three reasons why these natural—and, as we'll see later, "artificial"—toxic or potentially toxic substances pose no hazard under the usual conditions of consumption. First, the levels are usually so low, they have no effect. Of course if you eat huge amounts of a substance, you can do harm. For instance, too much cabbage can lead to an accumulation of goitrogenic substances, or an accumulation of lycopene following daily consumption of a half a gallon of tomato juice for several years can pose a hazard. But with reasonable diversity, no single chemical is likely to be consumed in hazardous amounts.

Second, individual chemicals appear not to be additive—that is, two or three different toxic substances in potatoes or lima beans or another source do not add up to make a hazard. Again, too much of one substance can.

Third, there is evidence that the toxicity of one element is often offset by the presence of an adequate amount of another. For example, the toxic effects of cadmium in a diet are reduced by accompanying high levels of zinc. (Perhaps this is what Mark Twain was referring to when he said, "Part of the secret of success in life is to eat what you like and let the food fight it out inside.")

In other words, we're back to the common-sense observation that the "safety in numbers" concept holds and that a well-rounded diet that excludes known hazards will contribute to, not undermine, health. The

---

* In other parts of the world lima beans have caused death.

rules are "be moderate about everything" and don't get caught up in the "if it's natural it's good" hoax.

So, one pillar of Healthfoodland has been shaken. What is natural need not necessarily be good—or even safe. But what about those "artificial chemicals"? What do they do and are *they* safe?

FIVE

What Have Additives Done for You Lately?

"The custom of saying grace at meals had, probably, its origin in the early times of the world, and the hunter state of man, when dinners were precarious things, and a full meal was something more than a common blessing."

CHARLES LAMB, Essays of Elia
(1823)

FOOD FOR THOUGHT

F O O D additives are like friends. We need and depend on them but often take them for granted.

We tend to take many of the positive elements of our advanced state of technology for granted—and even make them the subject of criticism. We routinely expect that we can come home at night after a day's work, walk into a well-stocked pantry, pick and choose exactly what we will have for dinner from the refrigerator, and have it ready in less than an hour. We naturally assume that the courses we prepare and eat will be nutritious, attractive, and enticing to our tastebuds. We expect that our meals will be enjoyable occasions, not just routine periods for satisfying biological needs.

But it wasn't always that way. And in parts of the developing world today, it still isn't that way. Historically, food supply and variation have been severely limited by the realities of season, weather, and food contamination and the food that was available most often provided a monotonous, unexciting menu. And as unappealing as it was, the meal probably took hours to prepare.

It is our variety of food additives and food-processing techniques that are responsible for making food as attractive, safe, enjoyable, varied, plentiful and, yes, nourishing as it is today. And it is food technology research that offers optimism for an even greater supply and variety for the future, and more nourishment.

But let's look back for a moment. We're hardly the first eaters to think about food additives and how they can make our lives a little bit more enjoyable and predictable. Although those with additive phobias may try to convince you that an interest in "food chemicals" is unique to those mad scientists born after 1950, point out to them that Epicurus, around the year 300 B.C., wrote of the joys of preserved cheese (once described as "milk's leap toward immortality"), that food colors were used in ancient Egypt, and that kerosene was burned in ancient China to ripen bananas and peas (the reason the method succeeded, although the Chinese did not know it, was that the combustion produced the ripening agents ethylene and propylene). Horace, back in 20 B.C., described as "the ultimate state of happiness" the existence of a food that would last for at least a year. And you can remind them that Marco Polo, Columbus, Cortes and others were sufficiently enthralled with food additives to take the risks that went with a long and dangerous ocean voyage to an unknown world. For Marco Polo and Columbus, the yearning was for spices. They wanted to supply the people back home with a means of vitalizing otherwise unexciting food. For Cortes, the prize was the vanilla bean, an additive that tasted so good that it made life in the sixteenth century truly worth living.

And if you need more evidence that additives are not twentieth-century inventions, there's always the fact that in 1691, some ambitious young enterpreneurs took out a patent for a means of "preserving by liquors or otherwise" and in 1783, another businessman applied and received a patent for preserving salmon with spices. Thus the additive business began by serving a need and by making life just a little bit easier.

Obviously the additives we use today go beyond the role of preserving cheese and salmon and offering the flavor of the vanilla bean. But what is it that additives do, besides provide the material for the scare books that try to convince us we are slowly eating ourselves to death?

THE CASE OF THE PURPLE PEACH
(AND OTHER CURIOUS TALES)*

Additives are substances intentionally added to food to enhance its visual appearance, taste, structure, storage

* This section presents only a brief overview of the role additives play in protecting and improving our food. If you want some detailed information on this subject, write to the Manufacturing Chemists Association, Inc., 1825 Connecticut Avenue N.W., Washington, D.C. 20009, or read *Handbook of Food Additives* by Thomas Furia, Chemical Rubber Company, 1968, or *The Chemicals We Eat* by Melvin Bernarde, American Heritage Press, 1971.

life, safety, or nutritional quality. (A skeptic of the "food-additives-are-poisons" philosophy might come up with another definition of an additive: "A pure substance of known composition added to a complex mixture of substances of unknown composition.")

Food additive critics claim that the benefits are generally very slight, that most additives are purely cosmetic and many involve outright deceit. But a closer look at the role additives play will reveal that these "consumerists" are overlooking a number of important facts.

Color Me Hungry

Consider first the additive function that is most under attack: coloring. Naturopaths say they don't care if their favorite fruit juice is perky purple or gangrenous green. Of course, that's up to them—and there certainly should be color-additive-free products available for them to choose. But there is another side to this very colorful issue.

Think about it for a moment. How would you feel about eating a bowl of purple peaches with yellow cream? Or how about some crispy toast with red butter, and a tall glass of chilled, green milk to wash it down?

It doesn't sound too appetizing, does it? At a recent meeting of one of the national associations of food color producers, the chef decided to make his meal cover the entire rainbow spectrum. And of course, he chose for his foods the colors that were least expected. When his guests got to the lilac-colored apple pie, dotted elegantly with black whipped cream, they began to wonder if the sandwich shop across the street was open.

It has been said that "we taste with our eyes." We can put up with some green icing on St. Patrick's Day and black and orange coloring on Halloween, but ordinarily we like our food to look "the way it should."

The expected appearance of a given food tends to reinforce its taste.

But wait, say food-color critics. We were *conditioned* to want it that way. If canned cherries and other fruits were always a drab beige in appearance we would never miss the bright, glistening red we associate with maraschino cherries. If we were to drop the use of all food colors now, our children would grow up not knowing the difference.

The situation, however, is just not that simple. First, food is a dynamic entity. Its color, as well as other of its aspects, changes with time. Frequently a change in color can occur although the food is still fully edible and highly nutritious. But it just doesn't look good enough to eat. For instance, there is much variation in the color of butter, both on a seasonal basis and in different localities. The addition of colorant assures uniformity. Similarly the flavedo, or outer layer of a ripe orange, frequently does not develop a characteristically orange color (see below). Food colors can anticipate and correct this situation.

Second, there is evidence that food must be appealing to be eaten. That's not a conditioned response. That's a gut reaction. This is especially true for children and sick adults who often need some type of incentive to eat. But it is true for the rest of the eating population too.

When oleomargarine was first introduced, it was illegal to include any type of coloring that would lead it to be confused with butter. Undoubtedly the butter industry was influential in getting this legislation passed. But, eventually, as an alternative, the oleo manufacturers were allowed to include a little packet of dye with their product—and the great majority of women would then faithfully mix this coloring into the white lardish mass to make it more appealing and more of an asset to their dinner tables.

But What Exactly Are All Those "Dyes"?

Any discussion of food colors must distinguish between (a) those naturally present in foods, (b) those of natural origin that may be added to foods, and (c) those of synthetic origin, the so-called "coal tar dyes."

It is obvious that nature contributed color to a wide range of foods, and many of these colors are used in their natural form—or in a synthesized form—to make our food more attractive. Carotenes, which are found in carrots, orange juice, apricots, tomatoes, and other foods, chlorophyll, found in leafy products, anthocyanins in berries and plums, are all useful in coloring a variety of foods (carotenes, for instance, are widely used in coloring oleomargarine and butter). There are other 100% natural forms of food coloring too, which we eat directly and as additives: saffron (the orange-colored aromatic, pungent dried stigmas of the species *Crocus sativus*), turmeric (the yellow, underground stem of *Curcuma longa*), and annatto ("natural orange," an extract obtained from the fruit shrub of the *Bixa orellana*). And while we are talking about 100% natural food colors, we can't forget cochineal, a natural red coloring that has been used for centuries. Unappealing though it sounds, this natural product consists of the dried bodies of the female insect *Coccus cacti*. These insects grow on a specific variety of cactus which is cultivated in the Canary Islands and parts of South America.

But for reasons of food production efficiency—and in the interest of keeping food prices lower—manufacturers since the mid-1800's have been using man-made colors in addition to those derived from plant and insect sources. The term "coal tar" to describe this classification of colors is derived from the fact that the early research in the dyestuff industries focused on various coal by-products for raw materials for coloring agents. Today, however, the materials for food coloring are directly synthesized in such a pure form that their original source could not be identified. It is, therefore, much more common to refer to these colors as "synthetic organic dyes" or more simply as "synthetic dyes."

Unlike the coloring agents from "natural" sources, synthetic dyes require certification, which means that each batch of dye must be inspected and approved as it comes from the manufacturer before it is used in food. The most widely used food coloring (it's found in soft drinks, ice cream, candy, baked goods, and elsewhere) is amaranth, better known as Red No. 2. Red No. 2 has been in widespread use in this country since the turn of the century without any reported hazard to human health. It has been more extensively tested than any other food coloring agent. But despite this long history of safe use, Red No. 2 was called into question in 1972 after Russian scientists suggested it might cause cancer or other health problems in animals. The FDA spent two years doing additional intensive testing on the coloring material and again concluded it was safe for use in human food. As described by *Food Chemical News* the "tests of Red 2 for carcinogenicity boiled down to two categories: one positive test and dozens of negative tests." But some consumer activists still complained. Perhaps what concerned them was the fact that Red No. 2 is chemically defined as

trisodium salt of 1- (4 Sulfo-1-naphthylazo)-2-naphthol-3, 6-disulfonic acid

"THANKS, BUT SINCE I HEARD THAT RED DYE IS A
COAL-TAR DERIVATIVE, I'VE BEEN OFF CHERRY SODA."

With a name like that, a coloring is handicapped right
off when being evaluated by those with "chemical pho-
bia." But the fact remains that Red No. 2 and all other
ten certified colors used in our foods today have been
carefully evaluated, carry with them specifications on
the maximum amount to be used and the conditions
under which they can be used, and are continuously
tested for possible problems. All current information
points to the fact that they present no hazards even if
you eat artificially colored foods in quantity every day.

In addition to the certified colors used in foods, the
FDA allows use of restricted coloring material, for instance, Citrus Red No. 2, which is used only to color the
skins of Florida oranges. As was mentioned above,
oranges do not always "look the way they should," and
the citrus coloring corrects for this. There is no current
evidence that this coloring treatment poses any threat

to human health—even when the orange rinds are eaten—but some people still worry. And some producers now offer their products "natural and green," although interestingly enough, they usually put in a word of explanation. One box of navel oranges arrived with the following disclaimer:

GREEN IS BEAUTIFUL...

Don't let my greenish tint bother you—inside I'm fully ripe. You see, my outside color depends on the climate. When it's cold I turn yellow-orange—the way you're used to seeing me. But this year, our nights have been mostly very mild, so my tree has been busily putting more natural sugar into me instead of color into my skin.

I'm not only *naturally* beautiful, I'm delicious!

(The *New Yorker* in noting this fruity explanation responded "You talk too much. Contemplate your navel!")

Over the years a number of health problems have been blamed on food additives. For instance, recently, Dr. Benjamin F. Feingold, formerly the Chief of the Allergy Department of the Kaiser-Permanente Medical Center in San Francisco, advanced a view that food additives were a major cause of hyperkinesis (a behavioral disorder which causes children to be overactive, impulsive, with short attention spans, and interferes with their learning ability). Specifically, he identified coloring and flavoring agents as the alleged villains. It is interesting to note here, however, that, to our knowledge, Dr. Feingold has never published one paper in the medical literature so that physicians and other scientists could evaluate his theories. Instead, he preferred to make his "findings" public through radio and television talk shows, press conferences, a popular book (*Why Your Child Is Hyperactive*)—and through at least one article in the Super Natural Rodale publication *Prevention*. Hardly the most professional approach to presenting scientific data for evaluation.

Do food additives make children hyperactive? Scientific advisory committees looking into this have concluded "no." The Advisory Committee convened by the Nutrition Foundation concluded "no controlled studies have demonstrated that hyperkinesis is related to the ingestion of food additives . . . the claim that hyperactive children improve significantly on a diet that is free . . . of food additives has not been confirmed."

But people were still concerned. And this panic was another example of what can follow the public announcement of random, unconfirmed conclusions. Shouldn't we have more than a few "testimonials" from just one researcher before we report something as a fact?

Shopping Once a Week

Dixon Lanier Merritt wrote, "A wonderful bird is the pelican. His bill can hold more than his belican. He can take in his beak food enough for a week. But I'm damned if I see how the helican." We differ from pelicans in that we have a daily need for food. If we had nothing else to do with our lives, we could spend the majority of our day shopping for food and preparing meals. Indeed, from the historical point of view the activities of a hunting and gathering society were focused in just this manner. But today we have other ideas about how we'd like to spend our time—and we thus need a stable, readily available food supply.

A major change in life-style accompanied the intro-

"Oh, drat! I forgot to add sodium propionate to retard spoilage!"

• •

duction of both efficient home refrigeration and the wide-spread use of commercial preservatives in food. You can shop on Friday, and still be confident that all but a few of the highly perishable foods will be safe and enjoyable until the following weekend.

Successful preservation of food isn't just a convenience. The World Health Organization estimates that 20–25 percent or more of the food supply is lost each year before it gets into consumer hands—as a result, for example, of infestation by pests and rodents and because of chemical deterioration. Prevention of chemical deterioration is one way additives prove useful.

Consider what happens when you peel a banana, peach, apple, or potato. If you don't get it in water pretty fast, it will turn dark (it's called enzymatic browning). A batch of potatoes that had been sitting, peeled, without benefit of water is, after its cooked, not the most appetizing of dishes.

For years, housewives have known about a simple technique of preventing vegetable and fruit browning: dipping each piece into the juice of a lemon, orange, or pineapple. The vitamin C in the juice prevents the change of color. That is, it acts as an antioxidant.

Antioxidants are added to a wide variety of foods to act as preservants. Fatty foods (for instance, margarine, cooking oils, some biscuits, salted nuts, and precooked dinners of fish, poultry, and meat) particularly need this type of preservative. Some antioxidants come from natural sources (citric acid, for instance). Others, including the familiar BHT and BHA are man-made products. In addition to the various types of antioxidants, additional types of preserving agents are used to inhibit the growth of mold, yeast, and bacteria, including "rope," a bacteria that can develop on bread.

OTHER GAMES ADDITIVES PLAY

"I can't eat it! It tastes terrible!" Familiar cries when new foods are introduced to youngsters. But they are right. Food has got to taste good, or people aren't going to eat it on a regular basis and in nutritionally correct amounts, no matter how healthy and nutritious it may be.

Art Buchwald in a 1960 issue of the *New York Herald Tribune* once complained about his experience with a liquid fad diet: "The powder is mixed with water and tastes just like powder mixed with water." Food has got to taste good if we are to get enthusiastic about it and derive pleasure as well as nutrition from a meal.

In an account by Sir Raymond Priestly of an expedition to Antarctica in 1910, we learn how a party was marooned for nine full months and suffered severely from the inhospitable climate and effects of starvation. About all they could find to eat was raw seal blubber.

It was either that or nothing. But it tasted so terrible, they couldn't get it down—until someone came up with a way of modifying its taste.

A somewhat parallel story revolves around scientists' attempts at the Institute of Nutrition of Central America and Panama (INCAP) to come up with an inexpensive high-protein product that would be useful in preventing kwashiorkor and other diseases of the malnourished prevalent in underdeveloped countries. They finally found one: a packaged combination of different vegetable products, including an oilseed meal and a cereal which together created a balanced protein food item (marketed under the name Incaparina). A miraculous finding. Except for one thing. The people for whom it was meant resisted eating it because it had an unacceptable taste. It just didn't have the flavor they expected a food to have. And further work had to be done to give the food more of a corn flavor to conform to the cultural pattern of flavor expectation.

Flavor is an important prerequisite to a nutritious diet. No matter how loaded with protein, minerals, and vitamins a food is, it will go uneaten if it doesn't please the palate. After all, eating is one of the pleasures of life.

There is more to eating than keeping a biological machine in operating order. We need an attractive diet. Dr. Melvin A. Bernarde in *Our Precarious Habitat* tells the story of a nutritional study in England made during the early stages of World War II when Britain, which relies heavily on imports to feed her population, was besieged. The study determined the nutritional components of an adequate diet and confirmed that a simple but tedious diet of green vegetables, bread, and milk could support vigorous activity. But the plan was not adopted, even though it would have reduced the need to import, because it was considered to be a far too monotonous routine for a highly advanced country. As British scientist Magnus Pike commented, "A diet may be perfectly balanced nutritionally, but if it is not sufficiently attractive a workman may not eat enough of it to do his work. If a chemist can enhance the attractiveness of such a diet harmlessly, he is in fact contributing to nutritional well-being."

So we need flavors, but where do we find them?

There are natural sources of flavor (and many of them are incorporated in our food supply today). But there just aren't enough. For instance, there is not enough natural vanilla in the whole world to flavor the ice cream we consume during a year in the United States. So we synthesize and from the laboratory come up with products that impart the same taste as the original version—but have the distinction of being available in virtually unlimited quantities. In some cases it is misleading to call these flavors "artificial" because they are *identical in chemical structure to nature's own.* They just happen to be made in the laboratory. And because the laboratory is their place of origin does not make them inherently sinister.

The "flavor category" makes up the largest single category of food additives. The estimate of the number of available natural and synthetic flavors ranges from 1,110 to 1,400. In addition to these direct flavors are the "flavor enhancers" such as MSG (monosodium glutamate) which do not directly alter the taste of food, but rather in some way affect the taste buds, perhaps making them more sensitive to picking up certain flavors.

Take a look in your kitchen and you'll find a few more of those "nasty" chemicals which make your life a little bit easier. Are you concerned about the emulsifiers in your salad dressing and chocolate milk? They're there to permit an even distribution of particles so that when you pour it or drink it, you're not treated to just one ingredient at a time.

One of the widely used emulsifier additives is lecithin, a derivative of corn and soybeans. There is something very ironic about a natural food enthusiast's condemning all "chemical food additives," which, of course, would have to include the emulsifier lecithin, while he is at the same time stuffing himself with the expensive lecithin wafers he bought in Healthfoodland.

You owe more to those much maligned chemical creatures than just cheerful, good-tasting, well-balanced food. The list of additive roles is long. Stabilizers and thickeners (including pectins, vegetable gums, and gelatins) give foods such as ice cream their uniform texture and desired consistency. (Can you imagine trying to eat an ice-cream cone filled with a substance that had the consistency of slightly thickened milk?) And there are additives that are necessary to control the acidity and alkalinity in making processed foods, added, for instance, to cheese spreads and processed foods to gain the desired texture and tartness.

Maturing and bleaching agents for wheat flour improve its baking results: freshly milled flour has yellowish pigments which are lacking in the type of quality necessary for elastic and stable dough. One way of getting around this problem—the manner in which flour was traditionally matured—was to let it age for about four months. The proteins in the pigments would oxidize and the flour would then be ready for baking. But that meant time and storage costs, and there was always the risk that the flour could be invaded by insects or rodents. Early in the nineteenth century it was found that certain chemicals could accelerate the maturing and bleaching processes. In the course of this treatment, nothing is removed from the flour nor are any residues left. It's just a matter of efficiency.

Various other food chemicals serve to retain a food's moisture, to give it a light texture, to maintain the clarity or firmness of a product, and to act as anticaking and meat-curing agents.

The term "processed foods" has recently raised concern among some consumers. Health food promoters describe this category of foods as "natural foods that have been adulterated in some way to deceive the consumer and make extra money for the manufacturer." Processed cheeses have been particularly under attack by consumer activists.

Before you panic, consider what "processing" really is. In the broadest sense *any* product that is treated in any way once it leaves the farm is processed. Pasteurizing milk makes it a processed product. And the processing of cheese is not much more complicated than that of milk. The manufacturer uses a variety of cheeses to ensure uniformity of flavor, heats them to the point of pasteurization, and pours them into molds to cool. Because natural cheese separates into its component parts when it is heated, an emulsifier, usually a type of phosphate, is added to keep the mixture in a homogenized state. Phosphates! Additives! But wait. Phosphorus is an essential element of all human life, animal and vegetable, and, despite what health faddists may say to the contrary, there is no reason to believe that it is anything but a health-promoting substance. Heating the cheese, adding phosphates so it stays mixed (not more than three percent phosphates are allowed by law), and cooling it in the desired shape—that's all cheese processing is about.

Occasionally some salt is used if the natural cheese base is lacking it, and possibly up to 1 percent water is used to achieve the desired consistency, but processed cheese is about as natural as you can get. And it has three distinct advantages over the type of cheese you get directly from the cheesemaker: it is uniform in taste and texture, it does not dry out or mold as other cheese does, and it does not always require refrigeration since it has already been pasteurized.

The use of additives gives us the opportunity to have a tremendous variety of foods. Compare the selection available at your local market and those on display in Healthfoodland. No wonder even the most dedicated of faddists "supplement" their diet with forbidden industrial goodies.

EAT YOUR ADDITIVES, THEY'RE GOOD FOR YOU!

Improving Upon Nature

Those who are alarmed about additives probably haven't taken the time to look at some of the ways additives have made our food healthier and safer. They tend to forget that fortification of food with vitamins C and D and iodide has all but eliminated scurvy, rickets, and goiter. When your antiadditive friend complains that processed bread has been "stripped" of its natural goodness, point to the enrichment it has received. And if he tries to convince you that unpasteurized "natural" milk is the only type to drink, tell him you prefer yours skimmed, without germs that cause undulant fever or bovine tuberculosis, and with an extra dose of vitamin A equivalent to that in whole milk.

PREVENTING DEADLY DISEASES

There is a rare, food-borne disease that has a very high fatality rate. It develops after the victim eats food products that have been contaminated by any one of a specific group of spore-forming bacteria. Symptoms of the disease usually occur between twelve and thirty-six hours after eating the poisoned food—nausea, vomiting, paralysis of various muscles, eventually double vision, drooping eyelids, dilated eyes, and, very often, death.

This disease is botulism.

The causative agent in botulism is a heat-resistant bacteria that will not be killed even at normal boiling points of water—the heating process must reach a higher level to eliminate these deadly spores. Many nonacidic foods are susceptible to the development of botulism, particularly some canned vegetables (thus the frequent mushroom recalls), which, because of incomplete processing, have not had all the spores eliminated.

As of this point, the best means of ensuring that the botulism agent does not develop on smoked meats is through use of nitrates and nitrites. Additives!

ADDITIVES THAT PREVENT CANCER

The widely used antioxidants BHT and BHA have been frequently criticized and labeled "health hazards." In the early 1960's, for instance, there were lurid reports to the effect that BHT caused eyeless offspring in rats. This finding was not confirmed in other studies performed around the world. Another researcher claimed that these preserving agents, when fed to pregnant laboratory animals, brought about brain changes and resulted in abnormal behavior patterns in rats. Again, these findings were highly questionable and the relationship of massive doses of BHT and BHA in rodents to the use of minute amounts of it in the human diet were not clear.

Recently, however, one United States senator called attention to these experiments, and stated that England had decided, on the basis of these minimal results, to ban the use of BHT and to restrict BHA. The pressure was then on to consider similar action in this country. In reply to the senator's statement, however, a high-level Great Britain health official said, "there's not a grain of truth" in the statement about a British ban. BHT and BHA are still being used in England and no change in their legal status is anticipated.

In the midst of all this controversy about the safety of BHA and BHT, some new information came to light.

These two antioxidants were first used in a widespread manner after 1947. They were added to cod-liver oil, margarine, lard, breakfast cereals, peanut, corn, and safflower oils, salted fish, bakery products, and among other things, butter. Testing continued through the 1950's and 1960's and then, surprise! Studies began to show that not only were BHT and BHA not harmful to health, *but it appeared that they had the capacity to inhibit the development of cancer.*

In laboratory experiments, it was confirmed that when BHT was added to the diets of mice, there was a marked reduction in stomach cancer. (It is hypothesized that perhaps these antioxidants prevent the attachment of the cancer-causing agent to DNA.) But that's in mice. Where was the evidence that it helped to prevent cancer in us?

Man does not lend himself well to direct experimentation. But there is what is known as a "natural" experiment, that is, looking back to see if any shifts in disease

Gastric Cancer Death Rates in Men in the United States, 1930–1968.

Age-Adjusted Death Rates Per 100,000 for Stomach Cancer 1964–1965

Country	Death rate M	Death rate F
United States	10.45	5.13
Canada	17.56	8.13
Australia	15.48	7.95
Austria	42.11	23.62
Bulgaria	40.56	26.67
Czechoslovakia	42.74	22.59
Denmark	21.76	13.39
England, Wales	23.42	11.46
Finland	39.66	20.38
France	21.44	10.63
Germany (F.R.)	37.09	20.69
Greece	16.49	10.04
Hungary	42.74	23.18
Ireland	23.88	15.94
Italy	33.61	17.81
Netherlands	28.26	15.18
New Zealand	16.54	8.33
Northern Ireland	21.87	13.59
Norway	26.01	14.63
Poland	44.18	21.17
Scotland	25.47	14.50
Sweden	22.04	12.03
Switzerland	26.04	14.90
Yugoslavia	21.10	11.95

rates can be identified and if so, whether they can be linked to any dietary change.

Between 1930 (when cancer statistics by site were first reported) and 1947, in the United States and in a number of countries with diets similar to ours, there was a relatively gradual decline in the incidence of stomach cancer. After 1948, however, there was a significantly more dramatic decline in this disease, while cancer of other sites (except of the lung) stabilized. Coupled with the evidence of the cancer-inhibition studies in animals, scientists now strongly suspect that the use of the antioxidants BHT and BHA played a role in that disease decline.

The United States currently has one of the lowest death rates from stomach cancer, followed by Australia, Canada, and New Zealand. The eating habits and methods of food preservation in these countries parallel ours. European countries such as Austria, Bulgaria, Czechoslovakia, Finland, Hungary, Italy, Poland, and West Germany, where there is no widespread use of antioxidants in food preservation, have the highest death rates from stomach cancer. Scientists have concluded that a small amount of "BHT certainly does not exert a harmful effect and may in fact be beneficial."

JUST TELL ME, ARE THEY SAFE?

So additives do play an important role in keeping our food supply plentiful, fresh, and attractive. And the use of some specific additives protects us from deadly diseases, perhaps even stomach cancer. But what about the majority of these strange-sounding chemicals—can we trust them?

First, you should remember, as you cross-examine the back of a salad dressing jar, that all foods are made up of chemicals. If the back of the "natural" milk carton read "contains lactose, casein, lactalbumin, calcium, phosphorus and more than 100 other chemicals," would *that* concern you?

Second, you should be aware that food additives, especially those that have been introduced in the last ten years, have undergone and have survived rigid testing procedures not applied to the great majority of natural products. And the standards for testing are very stringent. For example, before a food additive appears in your pantry, it has undergone between two and seven years of testing. The would-be manufacturer of the additive is required to perform acute toxicity tests to learn what the immediate effect of ingestion of the chemical has on a variety of laboratory animals. Next, the additive must survive a short-term toxicity test—usually about ninety days in length, involving different concen-

"Oh well, there goes another three million quid's worth of research."

trations of the chemical fed to several different animals. By means of constant monitoring, the animals are studied for general appearance, behavior, growth patterns, and eventually they are sacrificed and their organs dissected. Next comes the long-term toxicity analysis—usually covering two years or more—where scientists look for long-term effects and changes in fertility and reproductive behavior at various doses. Then and only then will the FDA consider a petition for approval of the additive.

The petition for approval must be accompanied by specified pertinent information, including full reports of investigations concerning the safety of the compound. The FDA then carefully evaluates all data, and when the scientific personnel are satisfied, then the product is declared "safe" for the intended use. To formalize this approval, a regulation is prepared and published in the Federal Register. The regulation for approval will specify, among other things, which foods the compound may be used with, the maximum quantities permitted, and the directions for labeling.

Of course if health problems were found at any of these stages, the whole project—and the one hundred thousand dollars (or more) worth of investment—goes down the drain.

If vitamin A were evaluated in the same manner, we'd have one less vitamin on our shelves.

Third, you should keep in mind that additives make up only a tiny fraction of the food we eat. Of the 1,500 pounds of food the average consumer eats in a year, just about nine or ten pounds are additives. (That is if you exclude the most common additives, sugar and salt.) Eighty percent (7.4 pounds) of this estimate is accounted for by about thirty of the most commonly used materials, of which about half are agents for leavening (e.g., yeasts, monocalcium-phosphate) and agents for controlling acidity and alkalinity (citric acid, sodium bicarbonate). The rest are the materials used for flavor (for example, mustard, pepper, MSG), propellants, carbonating and protective gases (carbon dioxide and nitrogen), and nutrient supplements (calcium salts and sodium caseinate).

Additives make up less than 1 percent of our total diet. It's been estimated that the median or "middle" level of use is one-half of a milligram of each additive per person a year (approximately the weight of one gram of table salt). In light of this, it seems logical that we should be worrying less about additives and more about the safety of the other 99 percent of the food we eat. Of course, if some additive or food is known to pose a hazard, then even "a little bit" should be kept out of our food supply. But we monitor our food constantly and have found no such hazard or any suggestion that various food chemicals are "stored up" for trouble later on.

Fourth, and finally, if your stomach begins to churn with anxiety when you read the words butylated hydroxyanisole, butylated hydroxytoluene and glycerol monostearate, you can calm it down by noting that all evidence points to the fact that the food additives we are using today are both safe and performing necessary functions. Our current food laws are such that an additive—especially one from a synthetic source—doesn't have a chance if it is found to pose even a slight health hazard to man or animal. A much practiced rule of thumb sets the acceptable percentage of the substance in the human diet at one hundredth of the no-adverse effect level in the animal—that is, a one-hundred-fold margin of safety. As a matter of fact, the laws are such (as will be pointed out in later chapters) that even some foods that haven't been given a full and fair trial may be removed from the market.

When in doubt about food additive safety, you can always try common sense. Let's say, for the sake of argument, that additives are harmful. Presumably then they would manifest their evilness in some way. But a study of disease statistics doesn't give much support to the additives-are-badditives theory: the only type of cancer that has occurred with greater frequency since the 1940's, when additives began growing in popularity, is lung cancer. And again, there is just no link here with additives. Heart disease is increasing, but no one has come up with a way to relate food additives to this problem either.

Dr. F. J. Roe, a British scientist, has pointed out that, if additives were dangerous, you'd expect to see a greater frequency of cancer of the gastrointestinal tract in developed countries. But it is just not so. As summarized by Dr. Roe: "It is clear that a lack of sophistication in food production and food processing methods is not an effective defense against the development of cancers of the gastrointestinal tract."

People who condemn food additives have issued a verdict without ever describing the nature of the crime.

We can't prove the ultimate safety of *anything*. Maybe that dentist in the 1940's was right. Maybe milk *does* cause cancer. We can only take a reasonable approach, based on our present knowledge about the advantages and disadvantages of a given food—whether it is "natural" or "artificial." And all our current studies indicate that food additives are safe. The consumer is in far greater danger from improper food preparation, storage, and plain overeating, than from food additives, given that their use is carefully regulated and revised when necessary. The very, very few instances of harm from excessive or careless use of additives, or from their unanticipated effects, are far outweighed by their many beneficial effects.

ADDITIVES VS. DRUGS

But despite the reassuring words of a large number of scientists, the use of chemicals in foods still raises eyebrows. It's really an emotional response. Some people just can't get out of their heads the image of a white-coated scientist emptying a bubbling test tube into their cream of chicken soup, and thus they are susceptible to health food ads which claim, among other things, that food additives are linked with everything from cancer to mental depression.

These same people will freely take drugs, and will accept the side effects that may accompany their use.

"It's just a reaction to all that artificial flavoring and artificial sweetening. Now if you just take this synthetic medication..."

© 1970 by *Medical Tribune*

A woman taking the Pill (which contains synthetic forms of estrogen and progesterone) every day for ten consecutive years may overlook the fact that this powerful drug has been shown to increase the user's chances of developing blood clots, gallstones, and other ailments.* But at the same time she will religiously avoid foods with a certain additive that has *never* been shown to present any health hazard to humans and she will adamantly demand that there be not even the slightest *trace* of the synthetic estrogen DES in the beef liver she eats once every two months.

The benefits of food additives are less apparent than those of drugs, not only to the consumer, but often to the health professional. But they are there and are sometimes even more dramatic than those of a "wonder drug."

FOOD REGULATION:
PAST AND PRESENT

All indications point to the conclusion that the food additives now in use are safe and contribute to good health. But as is the case with all food-related products, additives must be regulated and continually studied to ensure that those currently in food are being used in an appropriate manner and that new food additives have been fully and scientifically evaluated.

Man has long been concerned about the quality of his food and drink and the regulatory role now played by the FDA has a number of historical precedents.

Consider two random historical examples: one of the

* It has been estimated that about one woman in two thousand on the Pill each year suffers a blood-clotting disorder serious enough to warrant hospitalization. Pill users are twice as likely as non-pill users to undergo a gall bladder operation.

first food laws was the Assize of Bread, enacted in 1202 by King John. At that point consumers weren't worried about preservatives in bread, nor were they concerned with the white vs. whole wheat controversy. This first of all food laws strictly prohibited the use of such substances as peas or beans as substitutes for flour in bread.

In 1784, the first general food law was passed in the United States. Massachusetts enacted "an act against selling unwholesome provisions." Then, during the nineteenth century, various state and federal acts were passed, relating to inspection of food, tea, and other beverages, and to inspection of animals before slaughter.

Our modern food regulatory procedures began to develop in the late 1800's and early 1900's when there was a significant migration from country towns to cities, a greater sense of anonymity and expanded opportunities to profit from food adulteration. In 1906 Upton Sinclair, the advocate of periodic starvation as a means of self-purification, wrote *The Jungle* and drew attention to the dangerous filth that was accompanying the processing of meat and meat products in Chicago. The public was outraged by his findings and, as Sinclair put it, "I aimed at the public's heart and by accident I hit it in the stomach."

Dr. Harvey Washington Wiley, a chemist in the United States Department of Agriculture, was evidently among those who were hit in the stomach. He personally accepted the responsibility for monitoring the United States food supply and formed what was to be known as "the Wiley poison squad." Twelve healthy young men who pledged to eat nothing but what Dr. Wiley gave them served as guinea pigs in an ongoing experiment of the safety of the food their fellow countrymen and women were eating. If one of them showed signs of distress after eating a substance, that food was labeled "suspect" and consideration was given to withdrawing it from the market.

The Wiley system is similar to what we do today, except we don't use regular doses of food, we use overdoses, and we dissect our subjects (rodents) after a while to see if they've developed tumors in their innards.

Wiley and his associates set the stage for the food legislation we have today. Specifically, the poison squad recommendations prompted the passage of the Food and Drug Act of 1906, which required that chemicals be added only if they were safe and served a purpose. The act condemned "any added poisonous or other added deleterious ingredients which may render such article injurious to health." Loopholes in this legislation (for instance, one that led to the marketing of some deadly drugs) led to increased regulatory authority and eventually the passage of the federal Food, Drug and Cosmetic Act of 1938. But even after this 1938 act, the FDA was still in a position where it had to prove a drug or additive was harmful before it could be removed from the market—as opposed to having the manufacturers prove it safe before it was approved. Legislation during the 1950's reversed this situation and, according to the 1958 food additives amendment to the federal Food, Drug and Cosmetic Act of 1938, the burden of proof was to be on the manufacturer to prove the safety of a product *before* it received market clearance.

The passage of the 1958 amendment, and the new clearance authority of the FDA, presented a logistical problem. Would manufacturers be required to prove the safety of additives that had been in use for many years? Would salt and sugar have to undergo stringent testing? This seemed unreasonable, so, as an alternative, FDA officials divided food additives into three categories: (a) "prior sanctioned" substances, that is, those that had received official approval in earlier years and had a type of "grandfather status"; (b) "generally recognized as safe" items, the well-known GRAS list, which included some 700 substances, including ascorbic acid (vitamin C), caffeine, cinnamon, MSG, BHA, BHT, sugar, pepper, mustard—and cyclamates and saccharin, all of which had been approved by FDA chemists, toxicologists, and outside scientists; (c) all other ingredients not included in either (a) or (b).* It was this third category, made up of new substances or those that had lost their prior-sanctioned or GRAS status, that was to be subject to the closest regulatory scrutiny.

The food additives amendment of 1958 at first glance seemed pretty reasonable. The emphasis was on proving a substance safe—as opposed to proving that it wasn't harmful—before it was allowed on the American dinner table. It seemed to allow an ideal balance between legislative control and scientific discretion. That is, it did until something extra was added: the so-called Delaney clause.

The Delaney clause states that no food additive shall be permitted in the food supply if it is shown to induce cancer in man or animal. There are no ifs, ands, or buts about it. Once cancer is found in laboratory animals and is linked with the ingestion of an additive, regardless of the size of the dose or the length of time it was fed, that additive is found guilty and is sentenced to oblivion.

Even before it was passed, the Delaney clause pro-

* With the 1958 amendment, the term "food additive" took on a new and more limited meaning. For legal purposes, the "food additive" category *excluded* prior sanctioned and GRAS substances. In our discussion here, however, we use the term food additive in its more general sense.

voked more controversy than any other piece of food legislation in American history. This observation became even more prominent after it was recognized that the Delaney clause had laid the groundwork for the current fear of food additives and the related fascination with "100% natural, organic" products. It is a simple, straightforward, and totally inflexible law and it's worth taking a closer look at.

SIX

The Delaney Dilemma

> *"Law cannot stand aside from the social changes around it."*
> WILLIAM J. BRENNAN, JR.,
> *Roth* v. *U.S.* 354 U.S. 476 (1957)

FORTY-SEVEN WORDS

No additive shall be deemed to be safe if it is found to induce cancer when ingested by man or animal, or if it is found, after tests which are appropriate for the evaluation of the safety of food additives, to induce cancer in man or animals. . . .*

BACKGROUND

In 1958, as is the case now, cancer was the second leading cause of death. In 1958, as is the case now, cancer was a dread and not well understood disease.

---

* Food Additives Amendment of 1958, Section 409 (c)(3)(A), 21 U.S.C. Section 348 (c)(3)(A) (1964).

In the final hours of debate, Congress voted to include in the 1958 food additives amendment forty-seven carefully chosen words which would ban from the food supply any additive that had been shown to cause cancer in man or animals. Those testifying in favor of the bill (including actress and diet faddist Gloria Swanson) maintained that since so little was known about cancer, drastic measures should be taken to ensure that all food additives which were linked in any way with the development of this disease were immediately removed from the market.

According to Representative James J. Delaney (Democrat, 9th Congressional District of New York), congressional wives as well as Gloria Swanson played a major part in getting the legislation passed: "The members began hearing about our hearings from their wives. They'd buttonhole me in the hall and say, 'What's this all about? My wife has been giving me the devil about this bill of yours. She tells me that I've got to vote for it.'"

To those who wondered why just cancer (as opposed to other serious diseases that might be linked with additives) was the focus of the clause, Mr. Delaney explained: "Cancer is the most horrible of all diseases . . . What else would you single out?"

Of interest was the provision that the so-called Delaney anticancer clause did not apply to additives that were "previously sanctioned" or to those on the GRAS ("generally recognized as safe") list. Food chemicals in these special categories were not, in the strict legal sense, "food additives" and they were immune from the effects of the forty-seven words. Only those that for some reason had lost their GRAS or previously sanctioned status, or were introduced for the first time after 1958, were to be subject to the legislative rigors.

This apparent discrepancy, which focused on yet-to-be-introduced additives, led some people to wonder whether Congress favored old and familiar cancer-causing agents over newly discovered ones.

No American citizen in his right mind would advocate putting a cancer-causing agent in the food supply. From this point of view, then, the Delaney clause makes sense. As a matter of fact, looking at it from this vantage point, the clause appears unnecessarily redundant. If something causes cancer, no one would want to eat it and pressure would be put upon manufacturers to remove it from their products. So how could anyone complain about the Delaney clause? It seems to be as American as apple pie and no-calorie soft drinks. Or does it?

Before considering some of the more obvious limitations of the clause, consider first its alleged advantages; then review some of the rather dubious means in which it has been applied, and see why a number of people feel that the "anticancer" clause may be the primary underlying factor of the panic-in-the-pantry phenomenon.

DEFENDING DELANEY

Dr. G. Burroughs Mider, who in 1958 was the associate director of the National Cancer Institute, felt that the Delaney clause was just what was needed in the struggle against cancer. He pointed out that: "No one at this time can tell how much or how little of a carcinogen will be required to produce cancer in any human being, or how long it would take the cancer to develop." His argument was that a chemical which was linked, in even the most remote way, with cancer should be banned. There was always, he and others felt, a possibility of carcinogens accumulating in the body. Maybe a certain chemical had only a limited capacity to cause cancer, but when used with thousands of other chemicals with a similar capacity, it could add up to a real health hazard.

Today there are people of the same opinion. At a meeting of the New York Academy of Sciences in 1973, a group of scientists (although not necessarily a representative group) concluded that all attempts should be made to avoid exposure to food additives which have been shown to be even mildly carcinogenic in animals. These scientists went so far as to say that the spirit of the Delaney clause should be extended to include not only food chemicals, but also those in the air, water, and in other parts of the human environment.

James S. Turner (author of *The Chemical Feast*), writing in the *Vanderbilt Law Review*, agrees with this point of view. He calls the Delaney clause a "model environmental protection law" which he feels should be applied more widely. Turner claims that it is "model" in the sense that it "delegates to scientists the responsibility for making scientific judgments and to Congress the task of making policy decisions." In other words he is saying that the phrase "after tests which are appropriate" allows the scientists some leeway—but after they have reached the conclusion that a chemical is linked with cancerous development in laboratory animals, government regulatory activities automatically go into action.

There are others who defend the existence of the Delaney clause but do so in a manner that at least appears somewhat contradictory. Harvard's Dr. Jean Mayer, for instance, has stated that if the Delaney clause were first proposed now, he "would not vote for it, because if it was enforced literally, it would probably eliminate all foods." On the other hand, Dr. Mayer, who refers to the clause as a "sleeping watchdog," feels that now that it is law, he would rather not eliminate or alter it, but would prefer to see it "applied intelligently" by the FDA.

The Delaney Dilemma

DELANEY ON DELANEY

Representative James J. Delaney, a twelve-term congressman and the dominant figure behind the passage of the controversial anticancer clause, has become a hero in the fight for consumer rights. In his office in the Rayburn Building of the United States House of Representatives, he has a plaque* which reads:

NATIONAL HEALTH FEDERATION

HUMANITARIAN AWARD

to

Honorable James J. Delaney

for valiant efforts on behalf of his Countrymen's physical welfare, champion of the consumer's struggle for protective legislation against injurious additives in food and beverages

with admiration and gratitude this day
October 11, 1959

Fred J. Hart, President

* Despite its respectable sounding name, the National Health Foundation is not a medical or scientific group (and should not be confused with the National Health Council, a confederation of private and governmental agencies working for the public good in health matters). Instead, according to the magazine *Today's Health*, published by the American Medical Association, the Federation is a "rallying point for purveyors of food-faddist products, dubious and dangerous health information. . . ." *Today's Health* further reports that the Federation's leader, Fred J. Hart, who signed Representative Delaney's plaque, "is well known to federal authorities, and his name has appeared in the files of the AMA Department of Investigation for more than thirty years," and that Mr. Hart, whose background is radio operations and farming, has been in trouble with the FDA for distributing various "health" machines advertised as being effective in the treatment of a

141

In discussing his famous (or infamous) anticancer clause, Mr. Delaney recalls: "We worked the language back and forth, back and forth . . . on such a small clause as that we dotted every *i* and crossed every *t* . . . until we had something we thought could be understood by the public."

The word *ingest* was there by specific design, not by accident. Some of those involved were afraid that the clause could be used to ban anything that brought about cancer—whether or not the substance was eaten. One of those people who, back in 1958, expressed this concern was Elliot Richardson. He wrote:

> It would be important also to use language that would provide the intended safeguards without creating unintended and unnecessary complications. For example, the language suggested by some to bar carcinogenic additives would, if read literally, forbid the approval of use in food of any substance that causes any type of cancer in any test animal by any route of administration. . . . Concentrated sugar solutions, certain vegetable oils, and even cold water have been reported to cause a type of cancer at the site of injection when injected repeatedly by hypodermic needle . . . but scientists have not suggested that these same substances cause cancer when swallowed by mouth.

Representative Delaney did not stop his crusade for food safety after 1958. He is still fighting against "harmful" food additives, and he is particularly suspicious of the FDA efforts in this area: "There's nobody, so to

vast array of human ills, including cancer, coronary thrombosis, black widow spider bite, "discouraged" condition, rheumatism, loose toenails, and "confusion."

speak, on my side except some dedicated professor emeritus. . . . Food and Drug is supposed to protect the health of the public . . . you wonder what side they are on. They're with the manufacturers as far as I can see."

Conceding that he himself is not a scientist, Representative Delaney still maintans that "carcinogens are subtle, stealthy, sinister saboteurs of life. They have no place in our food chain." "Chemicals," he says, "do not have rights. People do."

Not only is Mr. Delaney suspicious of the FDA and its regulatory role, but he is suspect of any attempt "to improve on nature." He recalls his childhood when he would take a spoonful of Baker's chocolate, put it in hot water with some sugar—and it didn't mix. But today's products, he points out, have emulsifiers which are "so powerful [they] can tear down these molecules and make oil mix with water." He feels that this mixing process is a "great improvement from a commercial standpoint," but what is its effect on the body? On the liver? On the spleen?

I concluded that it must take a very strong chemical to bring about this result [the mixing]. What does such an emulsifier do to the body, I asked myself. . . . I figured that such chemicals must build up residual toxicity in the body. You know, some people can take six or seven drinks without showing any effects, and then with the eighth one they fall down drunk. It was this sort of thing that led me to become seriously interested in the problem.

Representative Delaney's forty-seven words in the 1958 food additives amendment were his way of making sure that if an emulsifier—or an artificial sweetener—could be found guilty of causing cancer (even on the basis of circumstantial evidence), it would be pulled from our pantry.

At first blush, it appears that the clause makes sense. Its intent is surely admirable—it clearly states that cancer-causing agents should be kept out of the food supply. It includes a provision which requires that the chemical in question be eaten (as opposed to being injected or implanted—or smoked) by the animal before it can be found guilty. And it seems to offer some scientific discretion by using the phrase "by tests which are appropriate." The whole clause generally sounds reasonable—until you take another look at it under cross-examination.

MORE QUESTIONS THAN ANSWERS

Seven critical questions, all of which point to the inflexibility and inconsistencies of the legislation, must be raised in evaluating its validity and practicality in modern-day food additive regulation.

First, the words in the Delaney clause leave no room for scientific discretion. If an additive causes cancer in animals, ban it. Of course, pro-Delaney individuals maintain exactly the opposite point of view. They claim that because the clause states that the test must be "appropriate," there *is* scientific adjustment permitted. But in reality, the atmosphere of fear and urgency created by the arbitrary nature of the clause precludes true scientific inquiry. If cancer is suspected, the Delaney flag is waved and the pressure is on. This will be evident in the case of cyclamates (described later in this chapter). The pressure was such that administrators were denied the "luxury of time." Because of the anticancer clause, the FDA is forced into bannings, not on the basis of any known or reasonably suspected hazard to human health, but rather because of the requirements of a very inflexible law. Food regulation becomes a matter of "cold figures" as opposed to in-depth, scientific consideration. And this approach doesn't allow for consideration of such realities as the *sponsorship* of the research. If General Motors sponsored a study that "proved" Ford cars were unsafe, would you accept its results as dogma? Of course not. But when the sugar industry sponsors a study to undermine cyclamates, Mr. Delaney's law does not ask questions.

Usually, you think of a rational scientific decision as being one based on facts. But "decisions" under the Delaney clause are derived from non-facts. That is, the condemnation of an additive, such as cyclamate, is based on a fear of what the sweetener *could possibly do* to human health, not what it has been shown to do—or even not on facts that would lead one reasonably to suspect a hazard. If we applied this principle to all foods, there would indeed be nothing left.

Identification of foods that may present health problems to man is a very complicated procedure, one that requires a heavy dose of scientific judgment and discretion. It is known, for instance, that rats fed low doses of a very powerful cancer-causing agent may well develop no tumors at all which can be linked with the cancer agent. But if another substance, in addition to the carcinogenic substance, is added to their diet (an oxidized fat fraction, for instance), all of them may develop tumors. And it could work the other way: that is, a substance shown to be cancer-related in one instance may prove harmless when combined with another chemical agent. The matter is just not as simple as it might initially appear to be.

"This serum has been found to be
very effective on white rats!"

Reprinted from *Medical Times*, January, 1971

Second, it is evident that there is no give and take in the Delaney clause, no room for a cost-benefit analysis. For instance, if nitrites were in the future shown to bring about cancer in ducks and were subsequently removed from their "previously sanctioned" category so they were subject to the Delaney clause, they could be removed from the market—even though nitrites were considered essential for preventing botulism. The inconsistency inherent in banning an additive that is performing a vital function, while acknowledging its presence in "natural substances" like spinach and some drinking water, is self-evident.

There is no leeway to take into account the benefits of an additive that protects health. As pointed out by Dr. Virgil Wodicka, "As things stand now . . . if it's found to induce cancer when ingested, you've had it."

The same analogy might be drawn for DDT. This

pesticide has saved more lives and prevented more disease than any chemical in history. Dr. Thomas H. Jukes, a professor at the Division of Medical Physics at the University of California, has noted that the life expectancy in India increased from thirty-two to forty-seven years during the 1953 to 1961 time period; this increase in longevity was attributed to the decline in malaria resulting from the DDT house-spraying program. But the induction of tumors in certain strains of mice by continuous dosage with DDT over a lifetime has been reported. And as of January 1, 1973, the use of DDT in the United States has been banned completely for most uses.

Third, the capacity of the Delaney clause to remove from the market any food additives that have been implicated in cancer development is inconsistent with the argument that many natural foods bring about cancer and are harmful, if not lethal, in many other ways. Why are we singling out just one category of potentially carcinogenic agents? Why, instead, aren't we equally concerned about *all* possible cancer-causing foods, not just those which happen to be arbitrarily listed in a certain column in an FDA publication?

The absurdity of the distinction between natural and artificial was recently raised again with regard to the regulation of the natural cancer-causing agent safrole in health food stores' sassafras tea. As was mentioned in an earlier chapter, safrole was recalled as a food additive in 1960, but just because it was nature's very own product, should it now be allowed in tea? The FDA didn't think so, and seized these products and a court case ensued. The tricky question raised here was where to draw the line, because many everyday spices have natural traces of safrole.

Fourth, there is the question of the relevance of the

experiments that find cancer in animals. If you suspend a couple of earthworms in ethyl alcohol for a few weeks, can you then draw conclusions which are applicable to the businessman who has two drinks before dinner every night? How can you feed a rodent amounts of cyclamate that would be equivalent to up to 1,300 bottles of diet soda a day, and then make generalizations to man?

In fairness, it should be pointed out here that animal studies do necessitate the use of higher levels of the test chemical than would be directly applicable to human exposure. Testing at high doses is essential to the experiment's attempt to identify in a relatively short period of time what is generally a low incidence of cancer development. But the question is, should an experiment be considered fair game for the Delaney clause if it uses *any* food additive in *any* amount for *any* length of time? Common sense tells you that too much of anything can harm you. Too much sun can cause skin cancer. But does that mean we should stay indoors all the time?

Fifth, it is known that cancer can develop in some animal species or strains, but not in another. As pointed out by Dr. Elizabeth Weisburger of the National Cancer Institute, the rats used in one experiment may have some type of bladder parasites that made them more susceptible to the development of bladder cancer. In one group of animals fed a strong known cancer-causing agent, no cancer will be noted (although other toxic effects probably will be), while in another strain 100 percent of the animals may be affected with cancer. But under the Delaney clause, cancer in one species or strain is enough to require legal action to remove the implicated additive from the market.

Drs. D. Mark Hegsted and Lynne Ausman of the Department of Nutrition of the Harvard School of Public Health have found that some commonly used laboratory animals—rats, for instance—can present unique problems as test subjects. Young rats, because of their very rapid growth rate, need much more protein than a young child. While a baby, for instance, can survive on human breast milk, young rats cannot survive on a diet of human milk. And the dietary needs of rats are different in other ways too: for example, these animals will grow normally without benefit of vitamin C or folic acid, while a human child will not.

Drs. Hegsted and Ausman question the relevance of some rat-feeding studies. One such study, for instance, concluded that an exclusive diet of eggs led to the growth and development of very healthy rats. But what is the relevance to man? Given the linkage of cholesterol-rich foods and heart disease, one would wonder about the validity of transferring these conclusions to man. In conducting studies of their own, these Harvard researchers pointed out that rats fed on an exclusive diet of fresh spinach do not survive for more than three days. Because of species-specific susceptibility—and different body metabolisms—it is not always valid to make generalizations.

Sixth, Dr. Weisburger also notes that "as the methods become more sensitive, they can find compounds where it was thought there were no traces of these materials. . . ."

If you have high-powered equipment, you can find just about anything in anything. As will become evident in the next chapter, this is what happened with the cattle-growth stimulant, DES. Although critics painted the FDA as being "the bad guy" for letting "harmful chemicals" infest our beef liver, it is clear that the detection of DES residues became evident only after supersensitive devices were introduced. As we get more sophisticated in our screening techniques, it is likely that this drawback of the Delaney clause will become even more prominent.

Seventh, those puzzled by the Delaney clause ask, "Why is it just limited to cancer? What about other diseases that might be linked to food?" Actually they ask these questions with some reluctance, lest they lend support to those who want to expand the clause. Senator Gaylord Nelson has introduced a bill that would broaden the Delaney clause (to plug another "gap in the food additive law") by stating that no food additive would be acceptable if it induced any biological injury or damage in any respect, including, but not limited to, the induction of cancer, congenital defects, and mutational changes.

Dr. Weisburger feels that the expansion of the Delaney clause would "open a bag of worms." Just about everything would go. Coffee, for instance, would immediately fall victim to this new law, as it has been shown to be mutagenic in lower species. Dr. Julius M. Coon, professor of pharmacology at the Thomas Jefferson University states:

> If such a bill became law, it would create utter chaos in toxicologic circles and in government regulatory agencies as well. It would be a real challenge, if not impossible, to find any additive that would not cause some harmful effects in test animals fed high levels over a long period of time. Strict interpretation of this law, for example, would quickly disqualify salt, sugar, vitamins A and D, and iron salts as food additives.

But the Delaney clause, with all its shortcomings, is still ruling our land. Take a look at the clause in action in two particularly infamous incidents (cyclamate and

DES), and consider how the mere existence of such a panic-inducing law is affecting our attitude about all "artificial" food products.

THE CYCLAMATE SAGA:
HOW SWEET IT WASN'T

If you asked the typical consumer prior to October 1969 what a cyclamate was, he or she might have told you that it was one of a pair of tandem bike riders. But after October 1969, cyclamate was a household word.

Contrary to popular opinion, cyclamates were not an original discovery of a mad scientist in the chemical-crazed 1960's. The sweetener was discovered by a graduate student at the University of Illinois in 1937.

Dr. Michael Sveda was doing some laboratory research when in the course of smoking a cigarette,* he brushed some loose tobacco shreds from his lips. Breaking one of the primary rules of laboratory protocol, Dr. Sveda decided to taste the chemical and later reported: "It was sweet enough to arouse my curiosity."

Dr. Sveda's transgression of laboratory rules led to the discovery of cyclamate. What is a cyclamate? Sodium and calcium cyclamates are salts of cyclamic acid (cyclo-hexysulphamic acid). But don't panic, read on.

When cyclamate products were first introduced in the early 1950's, they were intended for use only by those who had a medical reason to limit their sugar intake. But the word soon got around. Mothers saw cyclamates as an effective way of reducing both their children's sugar intake and hopefully their dentists' bills. Diet-conscious

* As a sidelight, it should be noted that Dr. Sveda, on the advice of his physician, has kicked the habit.

Americans saw cyclamate products as a means of satisfying their sweet tooth while still maintaining their waistline. Over the next eighteen years, cyclamates became extremely popular—carbonated beverages, fruit drinks, weight control preparations, canned fruits, jelly preserves, salad dressings, candy and baked goods could be offered on either a calorie-free or low calorie basis. People loved cyclamates. When they were mixed with a bit of saccharin (10 parts cyclamate to 1 part saccharin) they produced ideal sweetening results, free from the type of undesirable aftertaste that had been associated with the use of saccharin alone.

Between 1963 and 1969, the consumption of cyclamates soared from 5 million to 15 million pounds as Americans enjoyed the noncaloric sweet life.

It's Sweet, But Is It Safe?

Before cyclamate was introduced by Abbott Laboratories in 1951, it passed a series of systematic safety tests. It was assumed that the sweetener, after it was consumed, was eliminated from the body in its original form, that is, that cyclamate was not metabolized.

Early studies showed that excessive amounts of cyclamate could bring about a laxative effect. But it was felt, for practical purposes, a person using reasonable amounts of the sweetener had nothing to worry about. As summarized by the National Academy of Sciences' National Research Council:

> There is no evidence that use of the nonnutritive sweeteners, saccharin and cyclamate, for special dietary purposes is hazardous.

The Council did, however, recommend that the sweeteners be monitored in an ongoing manner in order to evaluate any possible long-term effects. As a result of this recommendation, and in light of the growing popularity of the sweeteners during the 1960's, both the producers of cyclamates and the FDA continued to monitor their safety. But test after test could reveal no health hazard associated with either cyclamates or saccharin.

Then on June 5, 1969, a University of Wisconsin study found that tumors developed in some white Swiss mice when cyclamate and cholesterol pellets were implanted in their bladders.

So, now that cancer was discovered, would the Delaney clause require that cyclamates be banned? Recalling the language of the anticancer clause, evaluators agreed that putting an implant of cyclamate in rodents wasn't the same as having them eat or drink it. The route of administration was inappropriate. So we could go on drinking our diet soda. For a while, anyway.

But Abbott Laboratories was concerned enough about the results of this study to initiate a new investigation. They asked the Food and Drug Research Laboratories (a private group in Maspeth, New York) to start *feeding* various amounts of a cyclamate-saccharin solution to rats.

Then the plot began to thicken. As mentioned before, it was generally assumed that cyclamate was released from the body in its original form, not metabolized and changed to another chemical. But in the late 1960's, it was found that in some people using cyclamate intestinal bacteria did internally convert small amounts of it to a substance known as CHA (cyclohexylamine). This finding raised some eyebrows, because it was felt that CHA might have an adverse effect of its own.

Because of this new interest in the cyclamate metabolite, the Abbott cyclamate feeding study, which was well underway, changed its protocol in mid-stream:

153

after the seventy-ninth week, in addition to being fed a 10:1 mixture of cyclamate and saccharin, half of the rats in each treated group were given dividend doses of CHA.

CYCLAMANIA

As of the early fall of 1969, after eighteen years of intensive research, no cancerous growths—or any other type of health problem—could be linked with the ingestion of cyclamate products. In October of 1969, however, an FDA scientist, Dr. Jacqueline Verrett, made a television appearance on the NBC evening news and described some gross malformations that had developed in baby chicks that had been injected with cyclamate. (Dr. Verrett now keeps her test results—chicks with oversized eyes, twisted skeletons, and other defects—in jars in her laboratory.) This scientist recommended that pregnant women not use artificial sweeteners lest their

'...AND WHEN YOU'VE FINISHED THAT, YOU'RE FIRED!'

© 1971 by *Medical Times*
Reprinted with permission of The Los Angeles Times Syndicate

offspring be exposed to the same serious type of problem.

She did not mention that injections of salt, water, or even air would have had similar or worse effects on the unfortunate chicks. Nor did the "informant" mention that all FDA regulations were being ignored in this disclosure of unverified conclusions. It is customary to first report data to professional colleagues for review and evaluation before announcing it to the world. But all the viewing public saw was some severely deformed chicks. And the association of the malformations and cyclamates was stressed.

Fear! Panic! Cyclamania! Some 75 percent of the American public were including cyclamate in their diet. And now it appeared that this sweet, innocent-looking white powder was actually a chemical villain.

Popular concern about the safety of cyclamate began to intensify. Then, on October 8, 1969, the results of the Abbott cyclamate-saccharin-CHA feeding study became available: of the 240 rats eating the cyclamate-saccharin mixture, 8 of those at the highest dose levels developed a form of bladder tumor; 7 of the tumors found were in animals that had been converting cyclamate to CHA. The addition of supplementary CHA midway in the experiment appeared to have no effect.

All the inevitable "experts" were called in—and all had their own, different opinion about what they saw. As a *Lancet* editorial described it, "Never have so many pathologists been summoned to opine on so few lesions from so humble a species as the laboratory rat." But eventually all did agree that at least four of the tumors were carcinomas. And the fate of the world's sweet tooth was thus based on the four unfortunate rodents.

On October 18, 1969, then HEW Secretary Robert H. Finch, announced that the findings of this single study required him to act in accordance with the Delaney clause and terminate the usage of cyclamate in food products. In Finch's words:

> I have acted under the provisions of . . . the so-called Delaney Amendment, enacted eleven years ago, which states that any food additive must be removed from the market if it has been shown to cause cancer when fed to humans or animals . . . because I am required to do so.

What appeared odd was that, in spite of the apparent urgency which prompted the banning decision, Mr. Finch allowed the condemned cyclamates to remain on the shelves for the next few months. The order did not become effective until January 1, 1970, and even then it was extended. Why, asked some confused consumers, if cyclamates posed a serious enough health threat to warrant such a sudden banning, were they allowed to remain at all?

In making the decision, Mr. Finch was putting aside the results of eighteen years of research that had demonstrated the safety of cyclamate. He was relying on a single, ambiguous study,* one that raised more questions than it answered.

First, since all animals received at least two compounds —and some three—why was just cyclamate banned? It is a basic research principle that "in studies using mixtures, it is essential that results be interpreted in light of available data on the individual components." Which of the three variables induced the bladder cancer? Cyclamate? Saccharin? CHA? Did the mixture exert some type of synergistic effect?

* Shortly after the ban, the FDA claimed to have its own study indicating that cyclamates were carcinogenic in animals—but the exact whereabouts of that study were, at the time, a source of confusion.

Second, could equally high doses of some other chemical (for instance, salt or sugar) bring about the same results? What about the possibility that it was the high doses, not the substances themselves, that caused a type of irritation on the bladder wall and the development of local tumors? (Five years after the ban, this explanation was accepted by many if not most scientists.) The rats developing cancer were being stuffed with the equivalent of up to 350 bottles of diet soda *every day*. It appears more likely that a human cyclamate freak following this intake pattern would drown before he had a chance to develop cancer.

"We Were Tried in the Press and Convicted"

Until October 18, 1969, many Americans had never even heard of the Delaney clause. And now here it was directly affecting their lives. Mr. Finch had said, in effect, "Delaney-Made-Me-Do-It." The real decision-making power was out of his hands and existed in a section of the 1958 food additives amendment.

But Peter Barton Hutt, general counsel of the Food and Drug Administration, feels that Secretary Finch was "dead wrong" with regard to his reference to the Delaney clause.

According to Hutt, the Delaney anticancer clause "had nothing to do with cyclamates." Of course, technically, he is right: cyclamate was an original member of the GRAS list. It was generally recognized as safe and thus was legally immune from the wrath of the Delaney clause. But from a more pragmatic point of view, even if the Delaney clause was not formally invoked, certainly its spirit was felt and in the last analysis it was the pressure of the Delaney clause that led to the sweetener's fall from GRAS.

WHY CYCLAMATES WERE BANNED

The decision-making process leading to the banning of cyclamate was a highly irregular one. Normal administrative channels were not followed and the decision was based on less than adequately established data. The precipitous banning of the sweetener was much more than just a necessary action, given the existence of the Delaney clause. Rather, it was the result of the interaction of four factors. Had the Delaney clause not been in effect, these four factors would not have had the impact they did.

First, there was the dramatic malformed chicken incident on a network television program. Lacking experience in biochemistry and experimental techniques, viewers were unable to evaluate the news reports, and just assumed that what they heard was accurate. According to former FDA Commissioner Dr. Herbert Ley:

> I would say . . . that inadvertent, probably self-serving efforts to bring this issue to the eager attention of the public as a topic for national debate on evening TV, did more to precipitate the rather impetuous actions on cyclamate than anything else. I would attribute these leaks to some scientists who were relatively low in the organization and who felt no esprit because the responsibility for regulation in foods was scattered throughout the agency at that time.

Or as summarized by another FDA official, "We were tried in the press and convicted."

Second, a cancerphobic American public was willing to ban a substance upon hearing the merest shred of evidence. Again, the cyclamanic reaction was not much different from that of the milk drinkers of the 1940's who panicked when Dr. Melvin Page informed them that milk was dangerous to their health. It is highly unlikely that cyclamate would have been banished as quickly as it was if there had not been a widespread, intense, and immediate specter of cancer. But indeed there was that fear, and combined with the Delaney clause, cyclamates didn't have a chance.

Third, the role of the sugar industry in the banning of the artificial sweetener cannot be overlooked. As a result of the increased use of cyclamate in soft drinks during the 1960's, the sugar industry suffered severe economic losses. The president of Sugar Information, Inc., and the head of the International Sugar Research Foundation in Washington, frankly admitted to the press that "sugar interests" backed more than half a million dollars worth of research on cyclamates since 1964. However, this research was done by responsible scientists in laboratories of good reputation. That some of the research was funded by the sugar associations should not necessarily make the results suspect, but it does suggest that the research should be verified before any action is taken.

On October 21, 1969, just a few days after the cyclamate ban, the *New York Times* noted that "Sugar Futures Make Sharp Gains," and indicated that there was very heavy sugar trading after the ban, since artificial sweeteners had been replacing some 700,000 tons of sugar each year in the United States.

Fourth, it appears that cyclamate was an easy victim of the Delaney clause spirit in that cyclamates had a willing and able understudy in the form of saccharin mixed with small amounts of sugar and a few alternative chemicals to diminish its bitter aftertaste. If no other sweeteners had been available, the decision might not have been made.

REACTION

The popular reaction to the banning of cyclamate generally was a mixed one—confusion with a bit of gratitude. Some consumers began to worry about the other food they were eating. After all, they had thought for years that cyclamate was safe, and now came all these frightening reports. What about those other chemicals with weird names? But there was gratitude as well as expressions of concern. A study of 343 students and weight watchers found that over 80 percent felt "gratitude for the government protection." The Long Island Press was even complimentary to the FDA:

> The federal action to end the use of cyclamates, a major component of artificial food sweeteners, is an encouraging sign of a growing sensitivity to the need for protecting man from some of his most ingenious chemical creations.

And the anti-industry, anti-FDA, pro health food segment of the population gloated over the whole situation:

> The battle was over and Finch had lost—along with the billion dollar industries that had been perfectly willing to keep on pouring a dangerous and treacherous chemical into its food and drink products, and the public be damned.

The reaction in professional publications was very different. Articles in Nature and Lancet and elsewhere criticized the FDA for setting off a bandwagon effect and acting "unscientifically," "irresponsibly," and "with a surprising lack of courtesy."

In retrospect, Dr. Virgil Wodicka agrees that the whole cyclamate matter could have been better handled: "We were not as experienced with this kind of exercise at this time and the handling was not exactly adroit."

By 1974, however, it began to appear that the cyclamate ban of 1969 was not only less than adroit, but also apparently unjustified. In 1973 the German Cancer Research Center in Heidelberg completed a series of cyclamate studies and reported finding no tumors that could be linked with the sweetener. In 1974, Abbott Laboratories submitted a sixteen-volume petition to the FDA, documenting the results of dozens of new studies which indicated that cyclamates did not cause cancer or in any other way interfere with health. Specifically, their reports indicated that cyclamates and cyclohexylamine (CHA) were found negative for carcinogenic potential in *twenty-two long-term studies in rats, mice and hamsters*—while only two rat studies suggested, on initial examination, any bladder carcinogenic potential for cyclamate. These latter two studies were the only tests *not* specifically designed for testing the carcinogenic potential of cyclamate and were further complicated by the presence of bladder parasites which may have caused the tumors. In addition, only four to sixteen rats in these two studies exhibited tumors, depending on the pathologist involved, compared to a total of 1,336 animals treated with cyclamates. Furthermore, the two studies with the positive results were considered by many—including FDA reviewers—to be lacking in recommended protocol for safety evaluation. In summary, Abbott had submitted material which was consistent with that of studies all over the world and which confirmed that cyclamates used as a sweetener were safe for human consumption. Indeed, it is likely that the Abbott documentation coupled with those of other researchers *made cyclamates the most thoroughly tested potential component of the human diet.*

The FDA pondered these materials for a number of months—and then told Abbott Laboratories that "our preliminary conclusion is that the petition does not contain the necessary information to support a finding that cyclamate is safe when used as an artificial sweetener . . . the carcinogenic potential of cyclamates is yet to be resolved." The agency told Abbott to go home and do more tests.

And the battle lines were drawn again. The *Washington Post* issued an editorial saying that "the FDA action is an example of how the agency should work . . . for now, cyclamate can rest in peace in the graveyard of chemicals." But Abbott did not have the graveyard in mind. Instead, at the time of this writing, it has its eye on the courtroom. If indeed Abbott does take the FDA to court, it will be the first time that the agency has been sued by a manufacturer over a food ban.

Why didn't the FDA act favorably on the Abbott petition? According to Abbott representatives, the FDA had told them off the record that even with the most convincing of data, the petition could not be approved unless there was a surge of public opinion in its favor. Apparently, at the time the FDA was making the decision, such a surge was not observed. But given the surge in *sugar* prices, the expressed need for some type of sweetener (saccharin is not only having its own legal troubles now, but does have the bitter aftertaste, and aspartame, a recently approved sweetening agent, is not a fully versatile one), *and* the possibility that an Abbott–FDA court case might bring the real facts to the attention of the public, there may well be a dramatic shift in public opinion.

Impact

Probably the single most significant implication of the cyclamate ban was its effect on public confidence about the safety of food additives in general. It was in October of 1969 that the panic-in-the-pantry syndrome took hold. President Nixon even acknowledged the problem in his October 1969 Consumer Message: he directed the FDA to begin a complete review (to the tune of 20 million dollars) of all food chemicals which were on the GRAS list—including salt and sugar.* All of a sudden, all foods were suspect.

As pointed out in a *Nature* editorial critical of the FDA ban of cyclamate, "It would be all too easy for public apprehension to be raised to the pitch where a fever of vegetarian faddism drives everything but mother's milk from the market."

And in a sense, this is the direction that public attitudes toward food additives began to take over the next five years. If there were no Delaney clause, and if the FDA and others had evaluated cyclamate in a rational, systematic way, it is highly unlikely that the sweetener would have been banned at all.

But more important, if the Delaney clause had not been exerting its force, it is unlikely that the general public anxiety about food additives would be as intense as it is now. Healthfoodland would be a remote hideaway for eccentric people, instead of the billion-dollar business it is today.

* At the time of this writing, the GRAS review is still in process.

THE BEEF OVER DES

How Now, Brown Cow

DES, the abbreviation for diethylstilbestrol, is a synthetic form of estrogen which has been used since the 1940's as a form of medication and since 1954 as a cattle growth stimulant. In 1973, DES was approved for "emergency" use as a "morning after" birth control pill.

Its impact on cattle growth is particularly significant: a 500-pound animal being treated with DES will reach a marketable weight of 1,000 pounds in 34 days less time, using 500 pounds less feed than would an animal not receiving DES. Additionally, it has been pointed out that DES causes an increase of seven percent in the amount

"You see, if you eat a great deal of beef, which is heavily injected with growth hormones..."

© 1974 by *Organic Food Marketing*

of protein and moisture in meat, while resulting in a decrease in the percentage of fat.* Prior to its banning in
1972, DES was being fed to seventy-five percent of the
30 million cattle slaughtered each year in the United
States.

It has been known since 1940 that DES, like all
types of estrogen, can bring about cancer in test animals
(breast cancer in mice and rats, cancer of the testicles
in mice). When the Delaney clause came into existence,
a special hole had to be punched in these forty-seven
words expressly for DES. Following the hole-punching
procedure, it became acceptable to use the drug as long
as no residue of it could be found in the meats. When in
1959 some residues were found in chickens, the permission to use it in poultry was withdrawn. But its use
in beef was still legal—again, providing that no residues
could be found. As a precautionary measure, cattle farm-
ers were required to withdraw their animals from the
drug forty-eight hours before slaughtering.

Testing for residues continued. But no DES was found
in the early 1960's. As late as 1968, however, the testing
techniques available were insensitive to DES traces at
levels below ten parts per billion (the same as about 5
drops per 25,000 gallons of liquid).

VAGINAL CANCER

In 1971, Dr. Arthur L. Herbst, a Boston gynecologist
associated with the Massachusetts General Hospital, published a paper in the *New England Journal of Medicine*,
reporting on young women who had developed an ex-

---

* The mechanism of action of growth-promoting agents like DES is
not clear and has been the subject of great debate. Suggestions include
a suppression of normal bacterial flora in the intestine, or an alteration
of the metabolic rate, relief from low-grade infections, protein-con-
serving effect, and in the case of hormones, the induction of a prolonged rapid growth akin to adolescence.

tremely rare form of vaginal cancer (adenocarcinoma). He had established that what these girls had in common was their mother's prenatal exposure to DES (a drug used previously to minimize the risk of miscarriage).

The link of cancer in these young women and the medicinal use of massive doses of DES added an emotional aspect to the evaluation of the cattle growth stimulant. By the end of 1971, the pressure was on and DES was under severe criticism. Cattle raisers were told to withdraw the drug seven days before slaughter, and testing for residues continued.

By 1972, a highly sensitive form of radioactive tracer technique was available, and DES residues were reported in about 2.5 percent of randomly selected animals. These traces, however, were only found in the *liver* of this small percentage of animals.

In July 1972, Senator Edward Kennedy held a subcommittee hearing on DES, stating, in a less than impartial manner, that "we are here today because DES, a known cancer-causing agent, is appearing on thousands of American dinner tables. . . ."

People were very concerned. They were panicked. They didn't want any "cancer-causing hormones" in their hamburgers (liverburgers?). The "us vs. them" attitude surfaced again. DES was described by critics as a means of "saving cattlemen some $90 million yearly." Of course, no link was made between the savings of cattlemen and the cost of ground chuck at the local market, which was to become the "prime beef" of consumers during the spring and summer of 1973.

INCONSISTENCIES IN THE BANNING OF DES

In the course of the beef over DES, little or no mention was made of some highly relevant points.

First, DES is an estrogen—but it is not the only source of estrogen to which American eaters are exposed. Milk, eggs, and honey have estrogen. It has been estimated that there is 1,000 times the amount of estrogen in an egg than there is in a serving of affected liver from an animal treated with DES. Additionally, the popular oral contraceptive has *far* more synthetic estrogen than would a serving of liver with DES traces.

Again, DES resembles all forms of estrogen and, like those from natural and "artificial" sources, can produce tumors in some susceptible strains of mice.

Second, and even more fundamentally, a woman's body regularly produces estrogen. *Nature* estimated that it would take 500 pounds of liver containing 2 parts per billion of DES to be equivalent (in terms of DES quantity) to the daily production of estrogen by a reproductive-age woman. Men, too, have their own share of natural estrogen.

Third, women who were formerly treated with estrogen to prevent miscarriage received up to 125 mg. a day during pregnancy. To ingest a dose equivalent to just 50 mg. of DES, you would have to, at one sitting, eat 25 tons of beef liver containing 2 parts per billion of DES. Today some women are using the "morning after pill" (as an emergency method). This medication contains DES. One would have to eat 1,000,000 pounds of liver from beef treated with DES to take in the amount of the synthetic estrogen contained in just one morning after pill.

Fourth, DES was condemned despite the fact that there wasn't any evidence that the minute residues found in the livers were harmful to man or animal. No one fed the liver containing the residues to laboratory animals and found cancer.

Fifth, all evidence points to the fact that there *is* a "no-effect" level for DES and other estrogens, that is, it is likely that relatively low amounts of the substance are not linked with tumor formation or any other health problem. This "no effect" principle has been demonstrated in mice—and the fact that all of us, male and female, have natural levels of estrogen, yet not all of us develop cancer, suggests that the same principle holds for humans. It *was* alarming to learn that massive doses of DES given to women during pregnancy could, apparently, have a cancer-causing effect on some of their female offspring, but such an observation should *not* lead us to be concerned about *traces* of estrogens in our food —especially given the existence of significant levels of natural estrogens.

But again, these are facts and expert opinion. And no food-panic fire will be put out with a whole bucket of facts and scientific judgments. As a result of the concern, and in light of the specter of the Delaney clause, on August 2, 1972, Food and Drug Commissioner Charles C. Edwards announced that he had "no choice" but to discontinue approval for DES in animal feed. At that time, he did allow the continued use of pellets (placed in the animal's ears) to release DES. But this exception didn't remain very long: in April 1973, the FDA also ordered the end of use of implants of DES, pointing out that studies had revealed 0.04 to 0.12 parts per billion (ppb) of DES in a sample of beef liver.

The manufacturers of DES did not take this action passively. After the FDA rejected their request for a hearing, they appealed the matter. On January 24, 1974, the United States Court of Appeals ruled that the FDA's orders were invalid and that the manufacturers of DES could resume marketing their products until the FDA had held a hearing. The court took the action because:

> The FDA chose to act summarily, without a hearing, without making known to petitioners the nature of the "new evidence" or of the underlying tests, and without giving the petitioners an opportunity to controvert the new evidence.

FDA counsel Hutt, as he did in the case of cyclamates, maintains that the Delaney clause was not invoked for DES. But no one will argue about the observation that it was an action inspired by the Delaney clause. Whatever the results of the subsequent DES hearings, it is clear that misinformed beef eaters will be unnecessarily concerned for years to come that there are "poisons" in their meat supply, put there for the benefit of "greedy cattle raisers."

The cyclamate and DES bans received a significant amount of publicity and drew attention to the severe limitations of the Delaney clause. But other additives have been affected by this piece of legislation, if only indirectly. The panic engendered by the inconsistent and precipitous actions with regard to the food we eat have made a large number of food products suspect.

SEVEN

Come and Get It
(Before They Ban It)!

> " 'Let the jury consider their verdict,' the
> King said, for about the twentieth time
> that day.
> " 'No, no!' said the Queen. 'Sentence
> first—verdict afterwards.'
> " 'Stuff and nonsense!' said Alice loudly.
> 'The idea of having the sentence first!' "
> LEWIS CARROLL, Alice's Adventures
> in Wonderland, CHAPTER 12,
> "ALICE'S EVIDENCE"

ADDITIVES ON TRIAL

A F T E R the banning of cyclamates, consumer groups began to raise questions about a number of the "food
chemicals" that find their way into our stomachs. All
that was necessary was one study, published or not, or
even a hint of a "suspicious observation" and someone
would propose a ban. Violet I, MSG, nitrates and nitrites, and saccharin became the subject of adverse press
reports, and some consumer stomachs began to churn
with renewed vigor. Others began to wonder if it wasn't
the Delaney clause, not the sweeteners and other additives, that should be on trial.

VIOLET I

Violet I was banned in April 1973, after some unpublished Japanese reports suggested it was carcinogenic in rats. Violet I was used by the Department of Agriculture in marking meat for grade and wholesomeness, and was used as a shading agent in other coloring procedures. Although the economic impact of this action had nowhere near the economic implications of the cyclamate ban, manufacturers of Violet I were less than pleased with the news. First, they couldn't understand why their product was being banned on the basis of a single piece of work by a single research team (the coloring agent had been used for twenty-two years). Second, they couldn't understand why the FDA refused to let them examine the results of the study at the time it was banned. Third, it was indeed puzzling (again) why after such precipitous and unchallengeable action, although its production was banned effective immediately, Violet I was allowed to remain in the products already on the shelves.

What's even more confusing is the fact that the FDA is currently retesting the coloring and an agency representative recently told the press, "If it comes out clear on our tests, then we'll bring it back." And consumers will be left to puzzle again over the question, "If it was considered so dangerous in 1973 that it was taken off our shelves, how could it possibly be allowed back?"

In the meantime, in place of Violet I a combination of three different red dyes (Reds No. 2, 3, and 40) and two different blue dyes (Blues No. 1 and 2) are mixed to give a purple color that will not fade on the meat over prolonged periods.

MSG

MSG, which, contrary to popular opinion, does not stand for Mighty Suspicious Goods, is monosodium glutamate. Ironically, MSG is a purely natural product, its place of origin being nature's own sugar beet. It is also found in seaweed, and it can be reproduced synthetically in a laboratory. It is simply the sodium salt of a common amino acid, a constituent of all protein, and is formed in the body whenever protein is eaten. Nothing could be more "natural."

The ancient Orientals were probably the first to enjoy the gourmet wonders of MSG. Exactly what MSG does, however, is not exactly clear. Presumably it does not alter the food; rather it heightens the sensitivity of taste buds and increases their response to specific flavors. One MSG advocate described its action as being like "turning up the volume on your hi-fi."

Others have suggested that MSG does what it does by increasing saliva formation. But whatever its action, many eaters feel a little MSG, "like a little love, surely helps." (Accent is MSG.)

But MSG, in all its natural glory, may create some problems. Or so say its critics. First there is the Chinese restaurant syndrome. In the late 1960's, a few people eating at Chinese restaurants began to complain about lightheadedness, burning sensations, and a type of facial pressure with a tightening of facial muscles. All those making reports were deemed to be sober, and it was concluded that the heavy dose of MSG in the Chinese cuisine might be responsible. Subsequent studies confirmed that some individuals may indeed have a sensitivity to MSG, just like some others may have a negative response to strawberries or tomatoes.

Second, there were some concerns which were potentially more alarming than the Chinese food syndrome. Dr. John W. Olney, an associate professor of psychiatry at the Washington University School of Medicine in St. Louis, administered massive subcutaneous injections to newborn mice and a single monkey and later noted brain damage. People quickly drew analogies to babies, a few consumer activists screamed for the removal of MSG from baby foods, and subsequently Gerber, Beech Nut, and Heinz voluntarily announced their intention to remove MSG from their baby foods.

Whether MSG is bad or possibly even good for babies is not the question. What disturbed many scientists was the panic reaction. On the basis of the mice data, the single monkey, and a protocol where the animals were injected, not fed, should we, asked some medical researchers, ban the *source* of MSG (the beet)? And what about the possible positive aspects of the use of MSG? Shouldn't we take them into consideration before we act? One study showed that under some circumstances, MSG can reduce blood cholesterol levels in man and certain animals. This observation and some of the findings about potential problems of MSG are very interesting, but at this point they are all inconclusive. We can't make judgments either way without verified facts. And we certainly can't condemn a substance that has been used and enjoyed for thousands of years on the basis of the mere hint of a problem.

The concern over MSG was most likely part of a fallout from the cyclamate incident. And although the Delaney clause did not apply here, because no form of cancer was involved, the same panic response was very prominent. Consumers were only moderately consoled by the findings of later MSG studies which indicated that at feeding levels of nine percent (a huge portion of any animal's diet), monkeys showed no adverse effects which could be attributed to MSG. It seemed that the earlier problems which preceded the voluntary omission of MSG in baby foods were due to the way MSG was administered to the animal (by injection) rather than to the additive itself. Nor were other antiadditive consumers pacified when they read the conclusion of the National Academy of Sciences to the effect that there was no evidence ". . . of hazards from the reasonable use of MSG in foods for older children and adults except for those who are individually sensitive to the substance."

NITRATES AND NITRITES

In 1972, an antiadditive enthusiast announced to all American housewives that bacon was the most dangerous thing in their refrigerators, since the nitrite treatment it received made it a potentially cancer-causing food.

Actually, this turned out to be less of a controversy than one might have expected—since prices at that time were so high that not very many people had bacon in their refrigerators. But they did have hotdogs and canned hams. So they were concerned. The specter of the Delaney clause was raised again.

Legend has it that by accident a caveman placed fresh meat on some sodium nitrate and found it kept for many days without spoiling. It also developed a pleasant taste and reddish color and today the nitrates and nitrites used in smoked and canned meats and in some poultry and fish products do exactly this. These substances give certain meat and fish products a fresh red appearance and produce the traditional cured flavor. Indeed it's impossible to have cured meats such as ham and bacon without nitrates because they are a part of the curing process. Nitrite, however, does more than just add color; most important, it prevents the growth of deadly botulism agents.

The safety questions that were raised stem from the possibility that nitrates may convert to nitrites, and the nitrites themselves may react with other substances (amines) to form nitrosamines. Experiments have shown that nitrosamines can cause cancer in animals.

The charges against nitrates and nitrites should, however, be put in perspective.

First, "natural" nitrates are commonly found in some water supplies and in spinach, beets, radishes, eggplant, celery, lettuce, collards, and turnip greens. These natural nitrates have the potential, under some circumstances, to convert internally to nitrite. This transition has been noted in infants—and has led to the previously mentioned warning about serving high nitrate vegetables to very young children. In the process of eating these natural foods, we take in far more nitrate than we get from cured meats and fish.

Second, nitrites themselves are also found in nature: the normal nitrite level of human saliva, for instance, is about 6 parts per million.

Third, there is no evidence at this time that nitrates and nitrites in cured poducts pose any hazard to human health. We do not know whether the low amounts of these substances that are permitted in food products actually combine with amines in the stomach to cause nitrosamines. What did cause some alarm was the observation that the U.S. Department of Agriculture had found traces of nitrosamines in 3 of 48 samples of processed meats studied. In tests conducted by the FDA, nitrosamines were found in one out of sixty hams tested.

It should be pointed out, however, that the levels of nitrosamines found in these samplings were very low— much lower than a level that would cause cancer in a laboratory animal.

It is known that massive doses of nitrites can be toxic in that they can interfere with hemoglobin, the oxygen-carrying material in the red blood cells. But this toxicity is hardly relevant given the minute amount of the substances used in meats and fish.

Fourth, nitrates and nitrites are added to foods to perform an essential and unique function: they prevent the development of deadly botulism-causing agents. At this time, there is no other known way to assure the safety of cured products.

Certainly nitrates and nitrites should be further studied. We don't know everything we should about how they work and what problems, if any, they might cause. The reaction brought about by nitrite in the curing process is very complex and not yet fully understood. It is known that the antibotulism effect of nitrite is accomplished by formation of some unknown compound early in the curing process. There is evidence that this effective compound is *not* a nitrosamine (e.g., not a known cancer-causing agent). Considerable attention is now focusing on the use of ascorbic acid (vitamin C) with the nitrates and nitrites. The ascorbic acid appears not to interfere with the protective effect of the nitrite— and indeed it may enhance it. Furthermore, the ascorbic acid seems to suppress the formation of the potentially dangerous nitrosamines.

Some people do not want to await the results of further studies of nitrates and nitrites. They are demanding that the Delaney clause be invoked immediately. Technically, the clause is not applicable here, because both nitrates and nitrites have a special type of immune status (nitrate is on the GRAS list, nitrite is "previously sanctioned," and under the law, neither of these categories is subject to the Delaney clause). By delisting nitrates and nitrites, they would become more vulnerable—but the situation is still complicated by the fact that nitrates or nitrites *used alone* might not cause cancer in animal experiments. The presence of some other chemicals (for instance, those in other common foods) might be necessary to bring about harmful effects.

The conclusion—at least for now—is that you can bring home the bacon (if you can afford it) and be confident that there is no hazard sufficiently great to cause alarm. There are those who demand their hotdogs with mustard and without nitrates and nitrites, but in the opinion of qualified scientists, the removal of these materials from our cured products would present a very definite health hazard.

SACCHARIN

Saccharin has a sweetening power 300–350 times that of sugar. Its two forms, sodium saccharin and calcium saccharin, are the salts of ortho-benzosulfimide. Discovered in 1879, saccharin was first challenged back in 1912 when Dr. Harvey Wiley of poison squad fame suggested it was not safe. The inevitable expert committee followed up his recommendations by banning the sweetener in food, but curiously enough allowed its continued presence in chewing tobacco. Obviously the ban on saccharin was eventually lifted. And when in 1958 the GRAS list was drawn up, saccharin was included.

Saccharin initially managed to maintain its status on the GRAS list despite its complicity in the experiment which led to the downfall of cyclamates. In January 1972, however, it was removed from the GRAS list and all products containing saccharin were required to carry a packaging disclaimer stating it was a "non-nutritive artificial sweetener for persons who must restrict their intake of ordinary sweets." At the same time it was recommended that adults use no more than one gram of saccharin per day, as if the ordinary soft drinker would know when he hit the one gram mark. (Actually, one gram is about equal to seven 12-oz. bottles of diet soda, or sixty of the small saccharin tablets.)

Soon after saccharin was removed from the GRAS list, a study financed by the International Sugar Research Foundation indicated that several rats in a group of twenty fed high doses of saccharin had developed bladder cancer. In February of 1973, the FDA disclosed that some of its own still incomplete studies suggested that animals fed high doses of saccharin developed tumors. In the FDA study, rats were fed saccharin for up to two years in doses from 0.01 percent to 7.5 percent of their total daily diet (7.5 percent in a rat's diet is roughly equivalent in humans to 1,300 bottles of a typical diet soft drink per day).

As of this writing, the future of saccharin is unclear. In early 1975, an expert committee, organized by the National Academy of Sciences' National Research Council, reviewed all the available data on toxicity studies that had been done on saccharin and concluded "the results of toxicity studies thus far reported have not established conclusively whether saccharin is or is not carcinogenic when administered orally to test animals." And with that, the committee ordered further tests. What is clear is that saccharin is being treated in a manner quite different from that afforded cyclamate in 1969. As summarized by

FDA's Donald A. Berreth, "We are a more mature agency."

The fact that saccharin is the only artificial sweetener left (aside from aspartame, which is not an all-purpose sweetener) surely puts it in a class all its own. There is a perceived need for some type of sugar substitute, so the prospect of the public upheaval over a banning may be enough to counterbalance the pressure from those demanding that the Delaney clause be invoked.

Several new artificial sweeteners are on the horizon and may someday prove to be competition for saccharin, which has that problem of a bitter aftertaste. But already questions are being raised with regard to them. For instance, aspartame (to be marketed by G. D. Searle Company under the name Equa) is a combination of two amino acids, L-aspartic acid and L-phenylalanine. A substance which weight for weight is about 180 times as

'This sugar substitute is perfect except for one thing.
It's salty.'

sweet as sugar, it received limited FDA approval in 1974. Aspartame has the drawback of breaking down under heat, cannot be successfully used in baking, and has not been approved for use in soft drinks. Like sugar, aspartame has about 4 calories per gram; saccharin has no calories. But within days after the announcement of this initial limited approval, Dr. John Olney, the same researcher who raised questions about MSG, was quoted in the press as claiming to have evidence that aspartame causes brain damage in children who eat it in conjunction with other additives. (He had MSG specifically in mind as the co-conspirator.) Again, this is a case of *one* scientist. But nevertheless, people were concerned about the sweetener before it even appeared on their grocery store shelves.

THE NATURAL STUFF IN YOUR SUGAR BOWL

Are they going to ban that too? Not quite. But people are concerned today about the safety of good old natural sugar. Specifically, there are rumors to the effect that sugar is the cause of diabetes and tooth decay, and a primary factor in coronary heart disease, obesity, and hypoglycemia. Additionally some people feel that sugar is not even a food or a nutrient. Are any of these accusations true? No. But food faddists—and those with personal economic interests in undermining confidence about sugar use—might try to convince you otherwise. It's worthwhile to deviate from our discussion of the Delaney clause and "artificial" additives to defend the case of old-fashioned sucrose.

Sugar is a pure digestible carbohydrate—and an important nutrient in our diet when used in moderation. Because it is a digestible carbohydrate, it is by definition a nutrient. Confusion arises here among some because the word nutrient is also used with a more restricted meaning to apply primarily to certain amino acids, fatty acids, vitamins, and minerals. These specific types of nutrients are occasionally set apart from other nutrients because some of them cannot be made from other materials in the body, and must be present in the diet. However, this is a semantic distinction only, and by no means implies that sugar and other carbohydrates are not nutrients. The fact remains that not only is sugar an important source of energy, but despite the sky-rocketing prices these days, it is the least expensive source in terms of our ability to produce calories agriculturally, and it is a most efficient product in terms of land use.

But what about the potential health hazards posed by sugar? First, even in excess amounts, sugar is *not* a cause of diabetes, although high levels of sugar consumption may exacerbate this disease. Second, for many people sugar may accelerate dental decay—but actually it is the sticky and excessive sugar consumption taken between meals which is involved in the acceleration of decay, not the sugar with meals.

And with regard to the relationship of sugar and coronary heart disease (CHD) there really is no basis for concern. Dr. John Yudkin and a few other researchers have suggested that sucrose is a causative factor here, but this view has been strongly criticized in the scientific literature (although the same view has been reported as "fact" in the popular press). There have been a number of scientific reviews of the data on sugar and heart disease, one of which concluded "although evidence is incomplete, such evidence as is available does not significantly incriminate sugar," and another reported "the theory is not supported by acceptable clinical, epidemiological theoretical or experimental evidence." Dr. Francisco Grande, a professor of physiological hygiene' at the School of Public Health of the University of Minnesota, reviewed all the "review articles" on this subject and wrote: "The weight of evidence seems to be against any direct association between high sucrose intake and the development of coronary heart disease." And when Dr. A. Stewart Truswell, a professor of nutrition and dietetics and head of the Food Science and Nutrition Department of Queen Elizabeth College, London University, was asked the question, "Does sucrose predispose an individual to CHD?" his response was "The experimental evidence for sucrose [being linked with CHD], unlike that for saturated fats, is unconvincing in the present state of our knowledge."

A number of private interest groups—the dairy and tobacco industries, for instance—would enjoy having some of the "blame" for heart disease put on sugar. But so far, the evidence does not point in that direction.

And a word about obesity and sugar. Obesity is caused by too many calories consumed and not enough expended in physical activity. It makes no difference whether the calories come from sugar, fat, or alcohol, too many are too many. Sugar is not always the culprit.

A few years ago, hypoglycemia (low blood sugar) was the "in thing," presumably the cause of most weakness, fatigue, and headaches. Of course sugar was thought by those who were "down" to be the cause. But Dr. T. S. Danowsky, professor of medicine at the University of Pittsburgh, writes that "very low blood sugars are all rare, exotic 'once in a million' events." And ". . . perfectly healthy persons can and do . . . develop a slight decrease in the blood sugar level without any associated symptoms or signs." Professor Danowsky concludes, "However, starch and sugars such as sucrose are not its [hypoglycemia's] cause."

WHERE DO WE GO FROM HERE?

The Delaney clause has, since its passage in 1958, been a source of sizzling debates among individual scientists—and between those representing "the public interest," who maintain they want to protect us from the environmental risks of cancer, and "industry," the "corporate devils," who are out to make money, even if it means we do all get poisoned.

Obviously, representatives of the food industry—or *anyone else* who depends on food for survival—want to keep cancer-causing agents off their dining room tables. But in the course of some highly emotional debates, people tend to overlook this obvious point.

And certainly we need a means of rationally monitoring all the foods and materials with which we come in contact. There is no doubt that on some occasions food and food-additive manufacturers *do* act in a manner which is not compatible with the public interest, and regulatory power becomes a must. But Mr. Delaney's forty-seven words—and the manner in which they have been applied—do not seem to be the answer. There must be a more reasonable way for the FDA to protect our health without leaving our cupboards bare and our sweet tooth decayed.

The louder the controversy becomes, and the more attention given to the FDA ambivalence about certain food additives, the more panicked consumers become. After all, scientists are supposed to know if a food is "safe" or "unsafe." That's their job. In many people's minds, there should be no cause for hesitation in classifying additives, that is, either endorsing their safety or banning them as a health hazard. When hesitation is noticed (as is the case with saccharin, MSG, nitrates and nitrites, various food colors), the conclusion drawn is that the additive is harmful.

CHEMICAL-PHOBIA

Public concern and the panic response are intensified when the additive in question is a "chemical" with an unpronounceable name. Possibly the flap over BHA and BHT is largely due to the fact that they have letter appellations, instead of "real names." One woman recently panicked when she read that her poodle's food contained "tocopherols." Chemicals! Fearing that her dog might be poisoned by this additive, she wrote the manufacturer, and was not relieved of her anxiety until she found out that those chemical "tocopherols" were actually the scientific classification for vitamin E.

Attempts at food regulation in the United States since 1969 have taken on a highly emotional hue. Food has always been highly susceptible to rumor. A few well-placed horror stories about any food product will set off a dietary alarm, and all the facts in the world won't dispel the feelings of fear which begin to mount. But the type of food legislation that guides us today further exacerbates the situation. By promoting precipitous decisions instead of rational on-going evaluation of food products, our legislation is lending support to those who make money from food fads.

CHEMICALS AND POLITICS

The feelings of panic engendered by the Delaney clause may very easily become intertwined with political factors. The FDA has three primary levels of activity: the routine operational activity level, the administrative guidance level, and the politically sensitive top level. Whether it's the sugar industry and its efforts to undermine cyclamates and saccharin, the dairy industry's attempts to implicate sugar in heart disease, or the influence exerted by a public interest group striving to achieve a "carcinogen-free environment," those on the sensitive super layer may, because of the arbitrary anticancer clause, be forced to overrule or misinterpret the rational scientific judgments that come from below. To some extent, this may be the case even without the clause. But the existence of the Delaney provision lends a certain type of urgency to the decision-making process, one that can easily be affected by political factors.

The FDA will always be in a difficult position in regulating foods. Any decision it makes will be unpopular with some group—either industry or the consumerists. The administration is now trying to alleviate the situation somewhat by changing its image from one of secretive paternalism, to one characterized by more openness. Ever since the clamor over the banning of cyclamates, the FDA has demonstrated an increased willingness to talk about its investigations and regulatory activities. To anticipate questions which are addressed to its personnel, the press office releases "talk papers" on subjects of current interest. In addition to this, there are frequent Food and Drug *HEW News* releases. And now the administration has regularly scheduled informal meetings, where members of the public are welcome to join in and ask whatever questions are on their minds. When, for instance, the national study of aflatoxins found that 25 percent of United States peanut products had "traces" of this cancer-causing agent, the FDA called

*"Good morning. I'm conducting a survey for
the Food and Drug Administration."*

a press conference to discuss the problem rationally. FDA's Donald A. Berreth feels that the failure to be open about a finding like this could have led to a dramatic headline, "FDA secret survey finds potent cancer-causing agent in 25 percent of peanut butter sandwiches." Instead, the press covered the survey in a reasonable manner, pointing to the FDA efforts to improve testing procedures so that eventually all aflatoxins might be removed completely from peanut products.

The public is concerned about the safety of food, and the staff of the FDA is concerned about their concern. In the course of all this turmoil over additives and their safety, consumers begin to think that the FDA is not doing its job. Just the opposite, however, is true. The efforts of the FDA have led to our food supply's being the safest and most plentiful in the world. The FDA *is* protecting our health and its members *do* want to alleviate the current unnecessary anxieties about additives. We have indeed been fortunate, in the past and currently, in having FDA commissioners who are intelligent, professionally trained, reasonable, dedicated, and apolitical individuals, the very type we should have in top regulatory roles.

They want people to know that they do have the health of the public, not the financial welfare of any individual interest group, as their top priority. But the existence of the Delaney clause makes these efforts particularly difficult. As one FDA representative put it, "The public . . . wants to be leveled with. But it's hardly an easy task to encourage the rational evaluation of the benefits and risks of food additives in an era of here-today-gone-tomorrow food legislation and the associated massive anxiety about 'those chemicals.' "

Pressure from "pro-consumerists," particularly Ralph

"You know what? I wish someone would recall Ralph Nader."

Reprinted through courtesy of *Medical Opinion*

Nader and his associates, also affects the nature of the FDA decision-making process—and has played a major role in the "return to nature" phenomenon. Referring to the recent flight from food additives to "healthy products," *Time* noted that the perfect recipe for a 100% natural business boom called for "the hip lifestyle . . . a pinch of nostalgia . . . and generous helpings of Ralph Nader."

It appears that there are certain individuals who have made a career out of condemning food additives and other aspects of modern life they see as "environmental hazards." Actually, the role of a critic is a very easy—although not always constructive—one to assume. When you add an intense public concern about cancer, an inflexible form of legislation, and a few scientists or pseudo-scientists who are prepared to challenge a food chemical at the drop of a rat, you have the perfect material for a panic-based, unscientific recall. When one researcher—no matter how preliminary or unscientific his or her data may be—can get front-page coverage and more attention than the results of years of meticulously conducted, replicated experiments, it is apparent that consumers are being seriously misled. When, for instance, FDA researcher Dr. Jacqueline Verrett went on national television in 1969 with her malformed chicks to suggest that cyclamates were not safe, eighteen years of careful research was challenged. Similarly in late 1974, this same researcher was featured on the front page of the *New York Times* and her "concerns" about the FDA approval of a food color (Red No. 2) were described in detail—despite the fact that systematic research at the FDA, involving a whole series of scientists, had given the dye "a clean bill of health."

In much the same manner, Dr. John Olney has led a

one-man struggle against MSG, one of the world's oldest additives, and more recently against the new sweetener aspartame. Questions about all foods we eat should be given serious attention, but when the requirements of stringent research protocols are finally met, the controversy in the press should cease, and individuals whose opinions are not consistent with the scientifically documented findings certainly do not merit having their names and views in headlines.

If any individual can perpetually sow seeds of doubt about food products and be given high priority attention, we will never enjoy a meal again. And if we continue to have laws that respond to this type of unwarranted pressure, we may soon find that our food supply—and variety—are very limited.

ON DELETING THE DELANEY CLAUSE

Agriculture Secretary Earl Butz calls the Delaney clause "a very unfortunate piece of legislation." Former Commissioner Edwards stated:

> My personal view and that of the FDA is that we [need] more flexibility of interpretation or we are put into the position that we were with cyclamates—all or nothing. And it becomes a highly emotional issue at that point, allowing no discretion on our part or anyone else's.

But others, including attorney James Turner, feel that Butz, Edwards, and other critics misunderstand the Delaney clause. It is Turner's contention that Delaney clause or not, cyclamates would have been banned and that, in a sense, the clause is an unnecessary duplication of already existing legislation.

FDA General Counsel Hutt agrees with Turner. Hutt calls the Delaney clause "the biggest red herring I ever heard of." He goes on to say:

> I have spent three years trying to convince the FDA that they shouldn't worry about Delaney one way or another. It's totally irrelevant in virtually everything we do.

But Hutt and Turner have not managed to convince food scientists—and consumers—of this. There is a strong feeling that even though the Delaney clause is only rarely invoked in its true legal sense, it still hangs there, like the sword of Damocles, preventing an un-panicked approach to evaluating the safety of our food supply.

What can be done about the controversial forty-seven words? How do we go about either diluting or deleting the Delaney clause? Counsel Hutt has one answer:

> It's very simple, what we'd do is we'd take off sugar, we'd take off all bacon and ham, we'd take off cyclamates, saccharin . . . salt, Vitamin A and five or six other things, we'd then play the Pontius Pilate type of routine, wash our hands and say "don't talk to us, go to Congress" . . . and the whole . . . public would get outraged and the law would be changed.

In other words, applying the Delaney clause in its literal, legal sense would quickly lead to its demise. But the likelihood of this type of super-ban does seem somewhat remote.

The first problem results from widely held assumptions

that (a) everyone is against cancer, (b) the Delaney clause is against cancer, and (c) anyone who is against the Delaney clause must be in favor of cancer. That's a politically unappealing label for an elected official.

Nobody is going to want to "come out for cancer." The first step, then, in refocusing food safety laws would be the reeducation of American eaters. Over the past five years panic books have been bombarding them with "facts" about how they are slowly eating themselves to a painful death. They are being told they are consuming lethal, cancer-causing food additives. Instead, more attention ought to be given to the real, *documented* facts, specifically, the observations that additives perform necessary health-promoting functions, are under constant surveillance, and are safer than many "natural" products. These statements are hardly as attention-getting as the "poisons-in-your-pantry" headlines, but they might precipitate a gradual shift from emotionalism to rationality.

The second problem in changing or eliminating the Delaney clause is the myth of "zero tolerance."

The zero-tolerance concept states that we should not tolerate any additive which brings about cancer—even though it is used in totally unrealistic amounts. The underlying premise of this concept is that we cannot establish any "tolerable" level of cancer-causing agent. But these zero tolerance claims do not always appear compatible with scientific reality. A mid-1950's study of an emulsifier (polyoxyethylene monostearate) used in baked goods and ice cream noted 13 bladder tumors in 174 rats who consumed a diet 25 percent of which was this emulsifier. This is an incredibly high percentage of one food in anyone's (or anything's) diet. It soon became evident to the scientists conducting this experiment that the tumors they were observing were directly

associated with the development of bladder stones, pro-
voked by the unrealistic levels of the chemical.

Later, numerous other studies, performed by a series
of different investigators, found that dietary levels of
2 percent, 5 percent and 10 percent of the same emulsifier
caused no bladder stones and no tumors. It was clear
that a "no effect" level of 10 percent could be considered
acceptable for the rats. The estimated maximum human
dietary consumption of this emulsifier was thought to be
about 0.025 percent. But, in a situation such as this, the
current Delaney clause would flatly deny the use of the
emulsifier in the food supply.

The zero tolerance principle today is nonsense. It
would make good sense if we were able to distinguish
absolutely by means of animal tests between substances
that caused cancer and those that did not, and if it were
possible to derive a diet that was truly free of naturally
occurring carcinogens. But neither of these "if's" is true.
So we find ourselves in a rather paradoxical situation,
banning so-called artificial substances (like cyclamates)
because they have been shown at extraordinarily high
doses to bring about cancer in rats, while tolerating a
natural substance (vitamin A, for instance), which under
the same protocol would also be classified as "cancer
causing." Additionally, we deny ourselves the option of
identifying a "no effect" level in animals and, at least
in some instances, using that information for making
decisions regarding human consumption.

U.S. Representative Clarence J. Brown of Ohio has
suggested that "a more logical approach would be to ban
only artificial substances whose carcinogenic properties
exceeded in amount of effects those . . . 'carcinogenic'
substances appearing naturally in our food supply." In
other words, under the Brown revision, cyclamates would
stay, as long as their cancer-causing effects were no

greater than those of vitamin A—or any other naturally occurring substance. Instead of worrying about whether 800 bottles of diet soda might bring about cancer, we could put our resources into making sure that *all the foods in our diet*—"natural" or "artificial"—are as safe as possible.

Another suggestion would revise the Delaney clause to read "if found to induce cancer when ingested by man or animals in amounts and under conditions reasonably related to its intended use." That would eliminate the 800-bottles-a-day-of-soft-drink problem.

But probably the most reasonable and rational suggestion for changing the food safety laws calls for the complete deletion of the Delaney clause. John C. Ayres, chairman of the Food Science Division of the College of Agriculture at the University of Georgia and Dr. Julius M. Coon, professor of pharmacology and chairman of the Department at Thomas Jefferson University, have succinctly stated their recommendation: *No revision of the Delaney clause would serve any useful purpose. It should be stricken from the law.*

Counsel Hutt has said that the Delaney clause would not be changed until there was something to replace it. But there *is* something to replace it: specifically, the Food, Drug, and Cosmetic Act of 1938 and the food additives admendent of 1958, of which the Delaney clause is just a small part.

The act and the amendment clearly state that poisonous or deleterious substances are not permitted in foods. The 1958 amendment strictly requires that there be premarketing clearance of chemical additives that are not on the GRAS list. Under these laws, no food that is even remotely linked with human cancer would be allowed in the food supply.

The existence of these general food laws, coupled

with the intensive review of GRAS list substances which is currently under way, makes the Delaney clause an irrelevant and unnecessary duplication of already existing authority.

There is no law in the drug regulation code comparable to the Delaney clause. Applications for new drugs are carefully and systematically evaluated, and the advantages and disadvantages of the new agent are considered. Most all drugs, if used in excess, could be linked with some disease, including cancer. But the FDA has worked out an elaborate framework for rational decision making about drugs, one which is consistent with our general laws about the safety of pharmaceutical products.

If the Delaney clause were dropped, the same rational form of evaluation of food additives would be encouraged. As it stands now, the clause is "a regulator's dream." Administrators are spared all the difficult decision-making processes. But scientific food regulation—as is the case with drug regulation—must be preceded by tough decision-making processes, ones in which the merits and drawbacks of each food or food additive are considered individually. No other country has a provision similar to the Delaney clause. Instead, each food additive is considered on the basis of its own unique attributes.

Congressman Delaney, in discussing his clause, recently stated that "it has saved millions of people from suffering." There is, however, no evidence whatsoever to support that claim. Indeed, the infamous clause may be responsible for *causing* a great deal of suffering—the type of anguish that turns what used to be a relaxing family dinner into a game of name-your-poison.

EIGHT

Regulating Our "Inner Environment"— Rationally!

"Fear cannot be banished, but it can be calm and without panic; and it can be mitigated by reason and evaluation."
VANNEVAR BUSH,
Modern Arms and Free Men (1949)

EATING—SAFELY—THROUGH THE EIGHTIES

F O O D and disease. We've always been concerned about the possible relationship between them. And it is highly likely that we will come to be even more concerned in the future. But the focus of that interest will not be the "artificial" vs. "natural" dichotomy. Rather it is probable that new information about both the toxicity and health-promoting effects of *all* types of food products—no matter what their origin—will give us a better understanding of the nature of the link of eating patterns and chronic disease. Our future in terms of food tech-

195

nology research in the late 1970's and 1980's will be focused on two related goals: the need to supply an adequate, healthy, and inexpensive food supply to a rapidly expanding population and the desire to develop the least hazardous, most health-promoting human diet.

FOOD IN A GROWING WORLD

The world population is now over the four billion mark, and is increasing at a rate that will cause the population to double about once every thirty-five years. Demographers have presented us with such staggering figures that we tend to block out the whole subject, but the reality of rapidly increasing humanity is with us.

We tend to forget that the preoccupation of most of the world is not with additives and their safety, but with getting enough food to live a healthy life. In global terms in the 1950–1970 period, with the widespread use of fertilizers and pesticides, food production increased at a faster rate than did population growth. (The world production of grains almost doubled while the population increased by less than 50 percent). But the situation has deteriorated in recent years, in large part because of droughts and general poor weather and higher birth rates. Starvation in many parts of the world is again becoming a leading cause of death, particularly among young children.

We need more food in the developed as well as the developing world. Intensive efforts are necessary both for increasing crop yield in each agricultural acre and for reducing food deterioration and wastage. In approaching this problem, we might consider three alternatives.

First, the return to "natural, organic, pesticide-and additive-free living," and the suppression of research that might develop new "chemicals" for food. Obviously, if we are now having problems in meeting food needs in developing countries and in being able to afford the food which is available in developed countries, abandoning the techniques of food technology that have taken generations to develop would only complicate the situation.

Second, we could look for new ways to protect and preserve food—for instance, techniques that do not involve the ingestion or digestion of food additives and pesticides. As an example, Dynapol in Palo Alto, California, is currently focusing its exclusive attention on the possibility of developing indigestible additives. Their primary concept is based upon the physiological fact that digestion is largely a process whereby food is broken down into its molecular components which are then reduced to molecules small enough to pass through the gut wall. Molecular size is, generally speaking, the only barrier to intestinal absorption. So if the additives could be altered in some way that would prevent absorption and still allow their functions to be performed, there is, at least theoretically, the possibility that current additives would not cause alarm among even the most dedicated faddists. But that's in the future, if at all. And of course there always remains the possibility that indigestible additives could have unique health problems of their own.

Third, we could move ahead in all aspects of food technology research, looking for new sources of foods, new and safe additives that could further reduce spoilage, and lead to the creation of new, highly nutritious, and inexpensive food supplements that would be culturally acceptable in areas where food supplies are limited. But a prerequisite of this type of rational approach is a trust in the "scientific system" and the acceptance of the fact that our current food research protocol has safety as its number one priority. To move in the direction of increasing and improving our food supply through scientific knowledge, we must have confidence in the food industry and in the regulatory agency that governs it. Right now, however, confidence in both seems to be waning.

Of course, this disenchantment with science is nothing new. We've long known that while technological abundance in any area initially elicits feelings of happiness and gratitude (when, for example, sanitary water, sewage disposal, and pasteurized milk became available, people responded positively; they could see the direct impact these public health measures were having on themselves and their families), the romance soon fades and is replaced by discontentment and suspicion. But we must return to the reality of life and acknowledge that we *do* owe a great deal to scientific advancements of the last seventy or so years—and that our hope for the future is based on the assumption that those advances will continue. Until, however, this feeling of confidence is reestablished, food research will be inhibited. Industries are understandably not eager to invest large revenues in projects that could be terminated by even a hint of a health hazard.

UNDERSTANDING CANCER

Also in the past seventy-five years, we have made major progress in the fight against acute, infectious diseases. The younger generations in the United States today are the first to grow up nearly free of the major threats of diseases such as whooping cough, scarlet fever, typhoid, pneumonia, streptococcal infections, and they know little about the serious nutritional diseases such as rickets and pellagra. But today we have another health problem:

chronic diseases and the quest for an understanding of those make up part of our challenge for the future.

We know that various foods—whether because they are naturally toxic or in some manner spoiled—can cause disease. And we also know that the lack of certain foods or an excess of food calories, saturated fats, cholesterol, can be the underlying or primary reason for illness. Until recently, however, in dealing with the relationship between causation and disease, we focused on the short-term, or acute, aspects of the relationship. We noticed that a food or the lack of a food caused an illness, we corrected the situation and health quickly improved. There was only a limited time between the cause, effect, and cure. But with chronic diseases such as heart disease and cancer, the time frame—and complexities of exposure—are such that relationships between cause and effect are not easily determined.

For instance, the oft-asked question is, what causes cancer? And the answer is that we don't really know— but we have good reason to believe that a variety of environmental (as well as hereditary) factors may be involved. This is a source of concern, in that we obviously want to avoid anything that has to do with cancer. But it shouldn't be the source of panic. What is it in our environment that might be suspect? It's been known for over two hundred years that chemicals we come in contact with play some type of role in the development of this disease.

An early description of the relationship of cancer and environmental factors was made by the English surgeon Percival Pott in 1775 when he noted an unusual incidence of cancer of the scrotum among chimney sweeps in England. It seems that the sweepers were exposed to the type of soot that tended to cling to the scrotal skin and eventually accumulated to the extent that it exerted a carcinogenic effect. After this eighteenth-century observation, the link of cancer with other external factors became more pronounced. We now know that bladder cancer occurs more frequently among workers involved with certain dye products, bone cancer is more frequent in employees who ingest radium, and lung cancer develops more often when industrial employees are exposed to, among other things, asbestos.

We also have reason to believe that various foods can increase the probability of specific forms of cancer. The epidemiological evidence related to aflatoxins in human liver cancer and laboratory data on such naturally occurring substances as safrole document this. We have learned that stomach cancer rates are unusually high in Japan and Iceland, very possibly because smoked meats and fish, which may contain the known cancer-causing agent benzo(a)pyrene, are eaten instead of chemically preserved foods. On the other hand, colon cancer rates are relatively low in Japan, again probably less because of some innate genetically determined resistance to this disease and more likely because Japanese eat less red meat.

We have a number of facts about food and cancer—and food and other diseases (for instance, saturated fats, cholesterol intake, and risk of various types of heart disease). And we are gathering more information each day. The evidence, however, does not always point to a cancer-induction effect. There are a number of hypotheses relating to the anticarcinogen effect of various foods. In other words, just as it is possible that certain natural foods or additives may be linked with cancer in the future, it is also possible that many of them may be shown to *reduce* the incidence of various forms of this disease. And, as we proceed in research, we may gather firm evidence that even for known cancer-causing agents the human body has a tolerance potential, that there is indeed a "zero effect" level of low doses of toxic but not hazardous materials.

We have to keep an open mind about all the food substances we ingest. We can't arbitrarily condemn one type and unquestioningly accept another. And in thinking about health and the possibility of food and other potential environmental hazards, let's not overlook the obvious. There is strong indication today that the American people are paying more attention to implied or uncertain risks than they are to large and unequivocal risks to health.

When we think about the relationship of chemicals and health—and before we panic about food additives—it's useful to remember that the "nastiest" and most lethal of chemicals are often used voluntarily by consumers. Tobacco. Alcohol used in excess. Inappropriate dietary composition. Illicit drugs. These are classifications of chemicals that are *known* to have an adverse effect on health. Just in terms of the first two examples, it has been estimated that in 1967, 75,753 cancer-related deaths and 185,359 cardiovascular-related deaths *were directly attributable to cigarette smoking*. In the same year nearly 60,000 deaths (from cirrhosis of the liver, cancer of the esophagus, alcohol-related accidents) *were attributed to alcohol abuse*. On the other hand, it has been estimated that more than 50 million of our current 210 million plus population would now be dead if the "chemicals" (both food additives and drugs) that were introduced after 1900 were *now not in use*.

TOWARD A HEALTHIER, BETTER-NOURISHED WORLD— WHERE EATING IS FUN AGAIN

We keep learning—about new sources and varieties of food, about diseases and their potential causes and treatment. And we're coming to accept the concept that nothing is "completely safe." There are only safe ways to use substances. We have a number of significant challenges to meet and many important questions to answer. So in conclusion, a number of specific recommendations may be appropriate.

First, it is clear that frequent panics about this and that substance "causing cancer" do nothing but slow down the scientific quest for facts. Every day, scientists all over the world come up with new findings, many of which are categorized as "preliminary." Scientific research is a slow, meticulous, repetitive process, and it is counterproductive to publicize "results" until they are verified, reviewed, and confirmed. Thus to see a newspaper headline which reads "Food and Health Experts Warn Against Bringing Home the Bacon" and an article which reports the results of *one* scientist who may or may not have data to back up his statements is not at all helpful to the public. Certainly consumers should be kept up to date on the results of laboratory findings, but the reports they read should be put in perspective—and indeed unverified scare reports based on the opinion of one or two scientists (or even someone with no scientific background) should be minimized. We need reporting and general public health education which gives a balanced account of the facts as they stand and helps reduce the communication barrier that now seems to exist between food scientists and consumers. Given that we all eat, we really do have a great deal in common regarding the question of food safety. The "us versus them" categorization is unrealistic.

Additionally we must have administrators who will not feel compelled to act on the basis of rumor. As summarized in a recommendation of the Report of the Panel on Chemicals and Health of the President's Science Advisory Committee (1973): "Where knowledge is so inadequate as to make the reality of a possible threat quite tenuous, the proper response is to seek more knowledge, not either to take drastic action or to do nothing." In other words we should neither panic nor look the other way if a question about safety is raised.

Second, we need food legislation that allows us to judge the things we eat not by whether they are "natural" or "artificial" or by what effects they may have when used in unrealistic quantities. Rather, we need laws that adjust themselves to advances in scientific knowledge and assess foods on the basis of their individual benefits, safety, and acceptability. Again quoting the Panel on Chemicals and Health: "The growing and changing nature of scientific knowledge demands flexibility in regulatory procedures—not rigidity. Laws, regulatory structures, and styles of administrative action all need to be adapted to a continuing growth and change in knowledge." In other words, we need food laws that will permit us to enter our pantries with a feeling of confidence, and a sense of assurance that what is there is safe, and what has been removed was banned for legitimate, scientific reasons, not because our laws could not keep up with advances in technology and scientific understanding.

Third, we must struggle more than ever against the deleterious effects of food faddism, the health food charlatans, and rumors about the hazards of specific foods. Right now we can't *afford* faddism. At best food fads are luxuries that are inconsistent with the worldwide food shortage and skyrocketing costs of eating. At worst, these fads are posing totally unnecessary risks to health.

We have to put our faith in the potential of scientific research to separate what is acceptable from what is hazardous, and assimilate the findings in a rational, unpanicked manner. Our alternative is a substantially limited, monotonous, expensive selection of foods on our shelves—or a continued feeling of uneasiness when we peer into our pantries.

And there is no reason for either of those two circumstances to prevail. Eating should be an enjoyable not an anxiety-provoking activity. That "all-softening, overpowering knell, the tocsin of the soul—the dinner bell" should promote pleasurable sensations, not a nervous stomach. So, in moderation, eat, drink, and be wary of those who raise questions about the safety of your food.

Selected References

"Abbott Labs Decides It Won't Withdraw Cyclamate Petition," *The Wall Street Journal*, October 10, 1974.

Abelson, P., "Chemicals and Cancer," *Science* 166:693, 1969.

Adams, R., "Natural Foods," *New England Journal of Medicine* 283:1058, 1970.

Adamson, R. H., *et al.*, "Occurrence of a Primary Liver Carcinoma in a Rhesus Monkey Fed Aflatoxin B1," *Journal of the National Cancer Institute* 50:549, 1973.

Alexander, T., "The Hysteria about Food Additives," *Fortune*, March 1972.

American Academy of Pediatrics, "The Use and Abuse of Vitamin A," *Pediatrics* 48:655, 1971.

Anderson, K., "After Cyclamates: What's Next on the FDA's Food Target List?" *Science Digest*, February 1970.

Annan, G. L., "An Exhibition of Books on Healing by Faith Fraud and Superstition in the 17th and 18th Centuries," *Bulletin of the Medical Library Association* 30:489, 1942.

Annino, L., *et al.*, "Non-Prescription Medications from Health Food Stores—A Potential Source of Serious Illness," *Connecticut Medicine* 35:428, 1971.

"Attorney General vs. Phony Health Food Claims," *Changing Times*, May 1973.

"Baby Food Additives May Be Next to Go," *Business Week*, October 25, 1969.

Bauer, W., *Health, Hygiene and Hooey*, New York: Bobbs-Merrill Company, 1938.

Beeuwkes, A., "Food Faddism and the Consumer," *Federal Proceedings* 13:785, 1954.

Bell, E. A., "Aminonitriles and Amino Acids Not Derived from Proteins," in *Toxicants Occurring Naturally in Foods*, Washington, D.C.: NAS, 1973.

Bernarde, M. A., *The Chemicals We Eat*, New York: American Heritage Press, 1971.

Bessman, S., and P. Hochstein, "Borscht, Beets and Glutamate," *New England Journal of Medicine* 282:812, 1970.

Bieri, J. G., "Effect of Excessive Vitamins C and E on Vitamin A Status," *American Journal of Clinical Nutrition* 26:382, 1973.

Bingham, E., "Thresholds in Cancer Induction," *Archives of Environmental Health* 22:692, 1971.

"Bitter Sweetener," *Time*, August 26, 1974.

Blackburn, J. E., "Report of the Proceedings of the Second Annual Convention of the National Pure Food and Drug Congress," 1899.

Bleiberg, R. M., "Cyclamate Ban: It Has Cost Producers and Consumers Dear," *Barron's*, October 7, 1974.

Blix, G., "Development and Features of Nutrition Fallacies in Sweden," in *Food Cultism and Nutrition Quackery*, G. Blix, ed., Uppsala: The Swedish Nutrition Foundation (printed by Almquist and Wiksells) 1970.

Borchert, P., *et al.*, "The Metabolism of Naturally Occurring Hepatocarcinogen Safrole," *Cancer Research* 33:575, March 1973.

Borlaug, N. E., in *Proceedings of the Western Hemisphere Nutrition Congress*, Chicago: American Medical Association, 1972.

Bové, F. J., *The Story of Ergot*, New York: S. Karger, 1970.

Briggs, M. H., "Fertility and High Dose Vitamin C," *Lancet*, November 10, 1973.

Brody, J. E. "Carcinogens: Unchecked, They Threaten an Epidemic," *New York Times*, October 6, 1974.

———. "Vitamin E Claims Held Misleading—Academy of Sciences Panel Warns Public on 'Cures,'" *New York Times*, September 12, 1973.

————. "Group of Scientists Warns Against Ban on Cancer-Causing Food Additives," *New York Times,* January 21, 1973.

Brooke, B., *Understanding Cancer,* New York: Holt, Rinehart and Winston, 1973.

Brooks, S., *The Cancer Story,* London: A. S. Barnes and Company, 1973.

Brown, C. S. "Cyclamates and Public Health," *Washington Post,* Sunday, September 29, 1974.

Brown, R. G., "Possible Problems of Large Intakes of Ascorbic Acid," *Journal of the American Medical Association* 224:1529, 1973.

Bruch, H. "The Allure of Food Cults and Nutrition Quackery," *Nutrition Reviews,* July, 1974, special supplement.

Bryan, G. T., and E. Erturk, "Production of Mouse Urinary Bladder Carcinomas by Sodium Cyclamate," *Science* 167:996, 1970.

————. and O. Yoshida, "Artificial Sweeteners as Urinary Bladder Carcinogens," *Archives of Environmental Health* 23:6, 1971.

Buckle, R., *et al.,* "Vitamin D Intoxication," *British Medical Journal,* July 22, 1972.

Butler, W. H., and J. M. Barnes, "Toxic Effects of Groundnut Meal Containing Aflatoxin to Rats and Guinea-pigs," *British Journal of Cancer* 17:699, 1964.

Butz, E. L., *This Week,* New Jersey Farm Bureau, July 10, 1971.

Calvin, M. E., *et al.,* "Salt Poisoning," *New England Journal of Medicine* 270:625, 1964.

Campbell, T. C., *et al.,* "Aflatoxin, M1 in Human Urine," *Nature* 227:403, 1970.

Carnaghan, R. B., "Hepatic Tumors in Ducks Fed a Low Level of Toxic Groundnut Meal," *Nature* 208:308, 1965.

"Carrot Juice Addiction Cited in Briton's Death," *New York Times,* February 17, 1974.

Carson, G., *Cornflake Crusade,* New York: Rinehart and Company, 1957.

Chaney, M., and M. Ross, *Nutrition*, New York: Houghton Mifflin Company, 1971.

Chemicals and Health, Report of the Panel on Chemicals and Health of the President's Science Advisory Committee, September 1973.

Chichester, D. F., and F. W. Tanner, "Nitrites and Nitrates," *Furia Handbook of Food Additives*. Cleveland: Chemical Rubber Company, 1968.

Christiansen, L., *et al.*, "Effect of Nitrite and Nitrate on Toxin Production," *Applied Microbiology* 25:357, 1973.

Citizens' Commission on Science, Law and the Food Supply, "A Report on Current Ethical Considerations in the Determination of Acceptable Risk with Regard to Food and Food Additives," January 28, 1974.

Cochrane, W. A., "Overnutrition in Prenatal and Neonatal Life: A Problem?" *Canadian Medical Association Journal* 93:893, 1965.

Cole, P., "Coffee Drinking and Cancer of the Lower Urinary Tract," *Lancet*, June 26, 1971.

Colloway, D., "Are Health Foods Worth It?" *McCalls*, October 1971.

Coon, J. "Natural Food Toxicants: A Perspective," *Nutritional Reviews* 11:321, 1974.

―――. "The Delaney Clause," *Preventive Medicine* 2:150, 1973.

Cooper, J. D., ed., *The Quality of Advice*, Volume 2, "Philosophy and Technology of Drug Assessment," Washington, D.C.: The Interdisciplinary Communication Associates, Inc., 1971.

Corry, C., *Quack Doctors Dissected*, London: Champante & Whitrow, 1810.

"The Costly Cyclamate Ban," *Norwich Bulletin*, December 2, 1974.

Crampton, R. F., "Problems of Food Additives, With Special Reference to Cyclamates," *British Medical Bulletin* 26:222, 1970.

Crawford, M., and S. Crawford, *What We Eat Today,* New York: Stein and Day, 1972.

Cummings, R. O., *The American and His Food,* Chicago: The University of Chicago Press, 1940.

"The Cyclamate Bandwagon," *Nature* 224:298, October 25, 1969.

"Cyclamate Ban's Time Has Come," *Hartford Courant,* November 26, 1974.

"Cyclamates: FDA to Re-Evaluate Banned Sweetener for Use as Food Additive," *Industry and Agriculture,* No. 30, February 12, 1974.

Cyclamates: Hearing Before the Ad Hoc Subcommittee of the Committee on the Judiciary, United States Senate. Ninety-Second Congress, Second Session on H.R. 13366, September 7, 8, 1972, Washington, D.C.: United States Government Printing Office, 1972.

Dahl, L. K., "Salt, Fat and Hypertension. The Japanese Experience," *Nutrition Review* 18:97, 1960.

Damon, G. E., "Primer on Food Additives," Department of Health, Education and Welfare Publication 74-2002, 1973.

Davis, T., *et al.,* "Excretion of Cyclohexylamine in Subjects Ingesting Sodium Cyclamate," *Toxicology and Applied Pharmacology* 15:106, 1969.

"Delaney Amendment," *Science News,* January 3, 1970.

"DES—Another Black Sheep?" *Nature* 238:67, 1972.

Deutsch, R., "Where You Should Be Shopping for Your Family," *Nutrition Reviews,* July, 1974, special supplement.

———. "Food Fads: Fantasy and Fact," in *Food and Fitness,* Blue Cross Association, 1973.

———. *The Family Guide to Better Food and Better Health,* Creative Home Library, 1971.

———. *Grass Lovers* (novel), New York: Doubleday, 1962.

———. *The Nuts among the Berries,* New York: Ballantine Books, 1961.

"Diet Industry Has a Hungry Look," *Business Week,* October 25, 1969.

"Donations for Agricultural Research," *California Agriculture*, March 1969.

Dougherty, P. H., "Advertising: On Health Cereals," *New York Times*, January 30, 1974.

Edelson, E., "Food Laws Should Be Even Tougher," *New Scientist*, February 8, 1973.

"Effects of Aflatoxin on the Liver," *Nutrition Reviews* 27:121, 1969.

Egeberg, R. O., et al., "Report to the Secretary of HEW from the Medical Advisory Group on Cyclamates," *Journal of the American Medical Association* 211:1358, 1970.

Elmund, C., et al., "Aflatoxins," *Journal of Chemical Education* 6:398, 1972.

Epstein, S., "The Delaney Amendment," *Preventive Medicine* 2:140, 1973.

Evans, I. A., and J. Mason, "Carcinogenic Activity of Bracken," *Nature* 208:913, 1965.

Evans, I., et al., "The Possible Human Hazard of the Naturally Occurring Bracken Carcinogen," *The Biochemistry Journal* 124:28, 1971.

"The Facts about Those So-Called Health Foods," *Good Housekeeping*, March 1972.

Fassett, D. W., "Oxalates," in *Toxicants Occurring Naturally in Foods*, Washington, D.C.: NAS, 1973.

FDA *Consumer Memo*, "Nitrates and Nitrites," undated.

FDA *July Fact Sheet*, Washington, D.C., 1971.

FDA *Talk Papers*, 1973–1974.

Feldman, M., and N. S. Schlezinger, "Benign Intracranial Hypertension Associated with Hypervitaminosis A," *Archives of Neurology* 22:1, 1970.

Finberg, L., et al., "Mass Accidental Salt Poisoning in Infancy," *Journal of the American Medical Association* 184:187, 1963.

Finch, HEW Secretary Robert, Press Release, October 18, 1969.

Fishbein, M., *Fads and Quackery in Healing*, New York: Blue Ribbon Books, 1932.

Selected References

Fitzhugh, O. G., *et al.*, "A Comparison of the Chronic Toxicities of Synthetic Sweetening Agents," *Journal of the American Pharmaceutical Association* 40:583, 1951.

"Food Additives: Blessing or Bane," *Time*, December 19, 1969.

Food Chemical News, 1970–1974.

Food and Drug Commissioner Charles C. Edwards, testimony before the Subcommittee on Intergovernmental Relations, House Committee on Government Operations, Washington, D.C., March 16. 1971.

Friedman, L., and A. T. Spiher, "Proving the Safety of Food Additives," *FDA Papers*, November 1971.

Fuller, J. G., *The Day of St. Anthony's Fire*, New York: Macmillan, 1968.

Furia, T. E., *Handbook of Food Additives*, Cleveland: The Chemical Rubber Company, 1968.

Fwii, T., and H. Nishimura, "Adverse Effects of Prolonged Administration of Caffeine on Rat Fetus," *Toxicology and Applied Pharmacology* 22:449, 1972.

Gardner, M., *Fads and Fallacies*, New York: Dover, 1957.

Gershoff, S. N., "The Formation of Urinary Stones," *Metabolism* 13:875, 1964.

Gibbons, B., "Cyclamate Come-Back? Court Battle Brewing," *Hartford Courant*, December 16, 1974.

———. "Controversial Cyclamate Sweeteners May Be Back," *The Times* (San Mateo), July 25, 1973.

Gittelson, N., "The Two Billion Dollar Health Food . . . Fraud?" *Harper's Bazaar*, November 1972.

Go, G., *et al.*, "Long Term Health Effects of Dietary Monosodium Glutamate," *Hawaii Medical Journal* 32:13, 1973.

Gold, G., "Egg Substitute Is Rated on Nutrition," *New York Times*, March 7, 1974.

Goodwin, R., "Chemical Additives in Food," Boston: Little, Brown and Co., 1965.

Graham, S., *Lectures on the Science of Human Life*, New York: Fowler and Wells, 1883.

Grantham, P., *et al.*, "Effects of the Antioxidant Butylated

Hydroxytoluene (BHT) on the Metabolism of the Carcinogens N-2-Fluorenylacetamide and N-Hydroxy-N-2-Fluorenylacetamide," *Food and Cosmetic Toxicology* 11:209, 1973.

Gray, D. W., "Battle over Sweeteners Turns Bitter," *New York Times*, June 1, 1969.

Green, R. C., "Nutmeg Poisoning," *Virginia Medical Monograph* 86:586, 1959.

Greene, W., "Guru of the Organic Food Cult," *New York Times Magazine*, June 6, 1971.

Hall, R. L., "Toxicants Occurring Naturally in Spices and Flavors," in *Toxicants Occurring Naturally in Foods*, Washington, D.C.: NAS, 1973.

———. "Food Additives," *Nutrition Today*, July/August 1973.

Hazelton, N., "Keep It Natural," *National Review*, August 18, 1972.

"Health Eaters," *Vogue*, May 7, 1967.

"Health Foods: Are They Nutritional Quackery?" *Good Housekeeping*, September 1967.

Hegsted, D., and L. Ausman, "Sole Foods and Some Not So Scientific Experiments," *Nutrition Today*, November–December 1973.

Henderson, L. M., "Programs to Combat Nutritional Quackery," *Nutritional Reviews*, July 1974, special supplement.

Herbst, A., *et al.*, "Adenocarcinoma of the Vagina: Association of Maternal Stilbestrol Therapy with Tumor Appearance in Young Women," *New England Journal of Medicine* 284:878, 1971.

HEW News Releases, 1968–1974.

Higginson, J., "The Geographical Pathology of Primary Liver Cancer," *Cancer Research* 23:1624, 1963.

Holbrook, S., *The Golden Age of Quackery*, New York: MacMillan, 1959.

Homburger, F., *et al.*, "Toxic and Possible Carcinogenic Effects of 4-Allyl-1, 2 Methylenedioxybenzene (safrole) in Rats on Deficient Diets," *Med. Exp.* (*Basel*) 4:1, 1961.

Howes, F. N., "Poisoning from Honey," *Food Manufacturer* 24:459, 1949.

Hutt, P. B., Letter to the Honorable Senators Gaylord Nelson, George McGovern, Alan Cranston and the Honorable Representatives Michael Harrington and Bob Eckhardt, Rockville, Maryland, July 9, 1973.

———. unpublished manuscript, dated April 13, 1972.

Inhorn, S. L., and L. F. Meisner, "Irresponsibility of the Cyclamate Ban," *Science* 167:1436, 1970.

Jameson, E., *The Natural History of Quackery*, London: Michael Joseph, 1961.

Johnson, A., and S. Wolfe, "Cancer Prevention and the Delaney Clause," Health Research Group, Washington, D.C., undated.

Jones, W. R., "Honey Poisoning," *Gleanings Bee Cult* 75:76, 1947.

Journal of the American Medical Association 120:1268, 1942.

Jukes, T. H., "Estrogens in Beefsteaks," *Journal of the American Medical Association*, Vol. 229:14, September 30, 1974.

———. "The Delaney 'Anti-Cancer' Clause," *Preventive Medicine* 2:133, 1973.

———. "Scientific Agriculture at the Crossroads," *Nutrition Today* 8:31, 1973.

———. "Fact and Fancy in Nutrition and Food Science," *Journal of the American Dietetic Association* 59:203, 1971.

Keen, P., and P. Martin, "Is Aflatoxin Carcinogenic in Man?" *Tropical Geographical Medicine* 23:44, 1971.

Kermode, G. O., "Food Additives," *Scientific American* 226:15, March 1972.

King, D., *Quackery Unmasked*, New York: S.S. & W. Wood, 1858.

King, R. A., "Vitamin E Therapy in Dupuytren's Contracture," *Journal of Bone and Joint Surgery* 31B:443, 1949.

Knipling, E. F., "Alternate Methods of Controlling Insect Pests," FDA Paper, Washington, D.C., February 1969.

"The Kosher of the Counter Culture," *Time*, November 16, 1970.

Kuhlmann, W., *et al.*, "The Mutagenic Action of Caffeine in Higher Organisms," *Cancer Research* 28:2375, 1969.

Lamden, M. P., "Dangers of Massive Vitamin C Intake," *New England Journal of Medicine* 284:336, 1970.

Little, L. C., and G. MacKinney, "The Color of Foods," *World Review of Nutrition and Dietetics* 14:59, 1972.

Mackinney, G., and A. C. Little. "The Coloring Matters of Food," *World Review of Nutrition and Dietetics* 14:85, 1972.

Maisel, A., "What Are the Facts about Food Additives?" *Reader's Digest*, May 1970.

Manufacturing Chemists Association, *Food Additives: Who Needs Them?* Washington, D.C., 1974.

———. *Food Additives: Everyday Facts*, Washington, D.C., 1973.

———. *What's Your Sign?* Washington, D.C., 1973.

Maple, S., *Magic, Medicine & Quackery*, London: Robert Hale, 1968.

Margolius, S., *Health Foods: Facts and Fakes*, New York: Walker and Company, 1973.

Mayer, J., "The Delaney Clause: A Sleeping Watchdog," *New York Daily News*, October 16, 1974.

———. *United States Nutrition Policies in the Seventies*, San Francisco: W. H. Freeman and Company, 1973.

———. *Human Nutrition*, Springfield, Ill.: Charles C. Thomas, 1972.

———. "Clinical Nutrition: Food Additives and Nutrition: A Primer," *Postgraduate Medicine* 46:195, 1969.

Melhourn, D. K., "Vitamin E: Who Needs It?" *Ohio State Medical Association* 69:751, 1973.

Miller, J., "Naturally Occurring Substances That Can Induce Tumors," in *Toxicants Occurring Naturally in Foods*, Washington: NAS, 1973.

Mintz, M., "The Culture of Bureaucracy: Rebuke at HEW," *The Washington Monthly*, December 1969.

"The Move to Eat Natural," *Life*, December 11, 1970.

Muenter, M., *et al.*, "Chronic Vitamin A Intoxication in Adults," *American Journal of Medicine* 50:129, 1970.

National Academy of Sciences, Washington, D.C., *Toxicants Occurring Naturally in Foods*, 1973.

———. *Interim Report on Non-Nutritive Sweeteners*, November, 1968.

———. *Toxicants Occurring Naturally in Foods*, Washington, Publication 1354, 1966.

Neary, J., "A Consumer Looks at the FDA As It Tries to Look Out for Him," *Life*, October 20, 1972.

Nelson, A. A., *et al.*, "Neurofibromas of Rat Ears Produced by Prolonged Feeding of Crude Ergot," *Cancer Research* 2:11, 1942.

"No New Life for Cyclamate," *Washington Post*, September 17, 1974.

Nutrition Foundation, Inc., *The Role of Nutrition Education in Combating Food Fads*, Symposium Report, 1959.

Nutritional Research Laboratories, "Lathyrism: A Preventable Paralysis," New Delhi, 1967.

Olney, J. W., "Brain Lesions, Obesity, and Other Disturbances in Mice Treated with Monosodium Glutamate," *Science* 164:719, 1969.

Olson, R., "Food Fadism—Why?" *Nutrition Reviews* 16:97, 1958.

Organic Foods, A Scientific Status Summary by the Institute of Food Technologists' Expert Panel on Food Safety and Nutrition and the Committee on Public Information, *Food Technology*, January 1974.

"Organic Shops Move into the Big Stores," *Business Week*, July 10, 1971.

Oser, B. L., *et al.*, "Conversion of Cyclamate to Cyclohexylamine in Rats," *Nature* 220:197, 1968.

Palmisano, P. A., "Vitamin D: A Reawakening," *Journal of the American Medical Association* 224:1526, 1973.

Pamukcu, A., *et al.*, "Lymphatic Leukemia and Pulmonary

Tumors in Female Swiss Mice Fed Bracken Fern," *Cancer Research* 32:1442, 1972.

———. "Assay of Fractions of Bracken Fern for Carcinogenic Activity," *Cancer Research* 30:902, 1970.

———. "Urinary Bladder Neoplasms Induced by Feeding Bracken Fern to Cows," *Cancer Research* 27:917, 1967.

Parham, E. S., "Attitudes Toward Banning Cyclamate," *Journal of the American Dietary Association* 56:524, 1970.

"Patulin, a Carcinogenic Myotoxin Found in Cider," *Nutrition Reviews* 32:55, 1974.

Patwardhan, V., and J. White, "Problems Associated with Particular Foods," in *Toxicants Occurring Naturally in Foods*, Washington, D.C.: NAS, 1973.

Pauling, L., "Ascorbic Acid and the Common Cold," *American Journal of Clinical Nutrition* 24:1294, 1971.

"Perils of Eating, American Style," *Time*, December 18, 1972.

Phillips, C., "Health Food Myths," *Vogue*, June 1967.

Phillips, W., "Naturally Occurring Nitrate and Nitrite in Foods in Relation to Infant Methaemoglobinaemia," *Food and Cosmetic Toxicology* 9:219, 1971.

Powell, H. B., *The Original Has This Signature: W. K. Kellogg*, Englewood Cliffs, N.J.: Prentice-Hall, 1956.

Price, J. M., *et al.*, "Bladder Tumors in Rats Fed Cyclohexylamine or High Doses of a Mixture of Cyclamate and Saccharin," *Science* 167:1131, 1970.

"Profitable Earth," *Time*, April 12, 1971.

Pyke, M., *Synthetic Food*, New York: St. Martin's Press, 1971.

———. "The Development of Food Myths," in *Food Cultism and Nutritional Quackery*, G. Blix, ed., Uppsala: The Swedish Nutrition Foundation (printed by Almquist and Wiksells) 1970.

"Regulation of Food Additives—Nitrates and Nitrites," 19th Report by the Committee on Government Operations, Washington, D.C., 1972.

Revelle, R., "Food and Population," *Scientific American,* September 1974.

Rice, W., "A Bittersweet Session on the Role of Sugar," *Washington Post,* October 27, 1974.

Richards, R. K., *et al.,* "Studies on Cyclamate Sodium (Sucaryl Sodium), a New Noncaloric Sweetening Agent," *Journal of the American Pharmaceutical Association* 40: 1951.

Richter, C., and J. Duke, "Cataracts Produced in Rats by Yogurt," *Science* 168:1372, 1970.

Roe, F. J., "Potential Carcinogenic Hazards in Foodstuffs," Proceedings of the *Royal Society of Medicine* 66:23, 1973.

Roe, F. J. C., and M. C. Lancaster, "Natural, Metallic and Other Substances as Carcinogens," *British Medical Bulletin* 20:127, 1964.

"The Role of Sugar in Modern Nutrition," Symposium held at Marabou, Sundbyberg, Sweden, August 18, 1973.

Rubini, M. E., "The Many Faceted Mystique of Monosodium Glutamate," *American Journal of Clinical Nutrition* 124:169, 1971.

Rynearson, E. H., "Americans Love Hogwash," *Nutrition Reviews,* July 1974, special supplement.

Saffiotti, U., "Comments on the Scientific Basis for the 'Delaney Clause,'" *Preventive Medicine* 2:125, 1973.

Sapeika, N., "Food Additives," *World Review of Nutrition and Dietetics* 16:334, 1973.

———. *Food Pharmacology,* Springfield, Ill.: Charles C. Thomas, 1969.

Schmeck, H. M., "Cyclamate Peril Denied by Maker," *New York Times,* November 14, 1974.

Schultz, D., "The Verdict on Vitamins," *Today's Health,* January 1974.

Scrimshaw, N., "Adapting Food Supplies and Processing Methods to Fit Nutritional Needs," in *World Population and Food Supplies,* 1980, American Society of Agronomy, 1965.

Sebrell, W., "Food Faddism and Public Health," *Federal Proceedings*, 13:780, 1954.

Selling, L. S., and M. A. Ferraro, *The Psychology of Diet and Nutrition*, New York: W. W. Norton, Inc., 1945.

Shamberger, R., and D. Frost, "Possible Inhibitory Effect of Selenium on Human Cancer," *Canadian Medical Association Journal* 100:682, 1969.

Shamberger, R. J., *et al.*, "Antioxidants in Cereals and in Food Preservatives and Declining Gastric Cancer Mortality," *Cleveland Clinic Quarterly* 39:119, 1972.

Sheraton, M., "A Skeptic's Guide to Health Food Stores," *New York Magazine*, May 8, 1972.

Shimkin, M., *Science and Cancer*, United States Department of Health, Education and Welfare, National Institutes of Health, 1969.

Silverberg, E., and A. Holleb, *Cancer Statistics*, 1972, American Cancer Society, 1972.

Smith, R., "Today's Health News," *Today's Health*, June 1972.

———. "A Phony Fountain of Youth," *Today's Health*, February 1966.

———. "Amazing Facts about a 'Crusade' That Can Hurt Your Health," *Today's Health*, October 1966.

———. "The Vitamin Healers," *Reporter Magazine*, December 16, 1965.

———. "The Bunk About Health Foods," *Today's Health*, October 1965.

Snider, A. J., "Beware Back-to-Nature Fads," *Science Digest* 72:44, 1972.

Spiher, A., "Food Additives," speech presented at the Richmond Dietetic Association, Richmond, Virginia, February 12, 1972.

Spodick, D., "Unsolicited Food Additives," *New England Journal of Medicine* 283:1413, 1970.

Stalvey, R., "Mr. Delaney Passes a Law," *Nutrition Today* 5:29, 1970.

Stare, F. J., "Current Nutrition Nonsense in the U.S.," in

Food Cultism and Nutrition Quackery. G. Blix, ed., Uppsala: The Swedish Nutrition Foundation (printed by Almquist and Wiksells) 1970.

――――. "Food Fads," *Today's Health,* March 1969.

――――. *et al.,* " 'Health' Foods: Definitions and Nutrient Values," *Journal of Nutrition Education* 4:94, Summer 1972.

――――. and M. McWilliams. *Nutrition for Good Health,* Fullerton, California: Plycon Press, 1974.

――――. and ――――. *Living Nutrition,* New York: John Wiley & Sons, 1973.

Stone, D., *et al.,* "Do Artificial Sweeteners Ingested in Pregnancy Affect the Offspring?" *Nature* 231:53, 1971.

"Sugar Substitutes Seek Sweet Smell of Success," *Chemical Week,* November 6, 1974.

"Sweeteners Await a Cyclamate Decision," *Business Week,* August 10, 1974.

"Sweetness and Light?: The Ban on Cyclamates Is a Triumph of Pseudo-Science," *Barron's,* November 17, 1969.

"Symposium on Additives and Residues in Human Foods," *American Journal of Clinical Nutrition* 9:259, 1961.

Tannahill, R., *Food in History,* New York: Stein and Day, 1973.

Tatkon, D., *The Great Vitamin Hoax,* New York: Macmillan, 1968.

Todhunter, E., "Food Habits, Food Faddism and Nutrition," *World Review of Nutrition and Dietetics* 16:286, 1973.

Trager, J., *The Big Fertile, Rumbling, Cast Iron, Growling, Aching, Unbuttoned Belly Book,* New York: Grossman, 1972.

――――. "Health Food," *Vogue,* January 1971.

Turk, A., *et al., Environmental Science,* Philadelphia: W. B. Saunders Company, 1974.

Turner, J., "The Delaney Anticancer Clause: A Model Environmental Protection Law," *Vanderbilt Law Review* 24:889, 1971.

Ulland, B., *et al.*, "Antioxidants and Carcinogenesis," *Food and Cosmetics Toxicology* 11:199, 1973.

Underwood, E. J., *Trace Elements in Human and Animal Nutrition*, New York: Academic Press, 1971.

Ungerer, M., "Make Mine Amino Acids," *Life*, May 22, 1970.

U.S. House of Representatives, Committee on Interstate and Foreign Commerce, A *Brief Legislative History of the Food, Drug, and Cosmetic Act*. Washington, D.C.: U.S. Government Printing Office, 1974.

———. *Vitamin, Mineral, and Diet Supplements*, Washington, D.C.: U.S. Government Printing Office, 1973.

VanDellen, T. R., "The Controversy over Vitamin E," *Illinois Medical Journal* 138:539, 1970.

Van Veen, A. G., "Toxic Properties of Certain Unusual Foods," in *Toxicants Occurring Naturally in Foods*, Washington, D.C.: NAS, 1973.

"Vitamin C for Colds" (editorial), *American Journal of Public Health* 61:649, 1971.

"Vitamins, Minerals and the FDA," *FDA Consumer*, DHEW Publication 74-2001, 1973.

Wade, N., "DES: A Case of Regulatory Abdication," *Science* 177:335, 1972.

Wagner, W., *The Golden Fleecer*, New York: Doubleday, 1966.

Wattenberg, L., "Studies of Poly-Cyclic Hydrocarbon Hydroxylases of the Intestine Possibly Related to Cancer," *Cancer* 28:99, 1971.

Weigley, E. S., "Food in the Days of the Declaration of Independence," in *Essays on History of Nutrition and Dietetics*, Chicago: American Dietetic Association, 1967.

Weil, A. T. "The Use of Nutmeg as a Psychotropic Agent," *Bulletin on Narcotics* 18:15, 1966.

Weiss, L., and E. D. Holyoke, "Some Effects of Hypervitaminosis on Metastasis of Spontaneous Breast Cancer in Mice," *Journal of the National Cancer Institute* 43:1045, 1969.

Wellford, H., "Behind the Meat Counter: The Fight over DES," *Atlantic*, October 1972.

Wen, Chi-Pang, *et al.*, "Effects of Dietary Supplementation of Monosodium Glutamate on Infant Monkeys, Weanling Rats and Suckling Mice," *American Journal of Clinical Nutrition* 26:803, 1973.

Whelan, E. M., "Fads vs. Facts: What's Your Food I.Q.?" *Reader's Digest*, March 1975.

———. "Can You Separate Food Fads from Facts?" *Glamour*, June 1974.

———. "How Sweet It Isn't," *National Review*, January 4, 1974.

———. and M. C. Quadland, *Human Reproduction and Family Planning: A Programmed Text*, Palo Alto, California: Syntex, 1973.

"Whiskey . . . with Charcoal, Please," *Pecos Enterprise*, January 3, 1972.

White, P. C., "Let's Talk about Food," *Today's Health*, February 1972.

"Why Cyclamates Were Banned," *Lancet*, May 23, 1970.

Wilson, B. J., and A. Hayes, "Microbial Toxins," in *Toxicants Occurring Naturally in Foods*, Washington, D.C.: NAS, 1973.

———. *et al.*, "Toxicity of Mold-Damaged Sweet Potatoes," *Nature* 227:521, 1970.

Wogan, G. N., "Naturally Occurring Carcinogens in Foods," International Symposium on Carcinogenesis and Carcinogen Testing," *Progress in Experimental Tumor Research*. 11:134, New York, Karger Basel, 1969.

———. and R. S. Pong, "Aflatoxins," *Annals of the New York Academy of Science* 174:623, 1970.

Wolff, I. A., and A. E. Wasserman, "Nitrates, Nitrites, Nitrosamines," *Science* 177:15, 1972.

Wolff, R., "Who Eats for Health," *American Journal of Clinical Nutrition* 26:438, 1973.

Wolnak, B., "Food Industry and FDA Face Food Fad

Threat," presentation before Food Update Eleven, April 17, 1972, Key Biscayne, Florida.

Yergin, D., "Supernutritionists," *New York Times Magazine*, May 20, 1973.

Young, J. H., "Historical Aspects of Food Cultism and Nutrional Quackery," in *Food Cultism and Nutrition Quackery*, G. Blix, ed., Uppsala: The Swedish Nutrition Foundation (printed by Almquist and Wiksells), 1970.

———. *The Medical Messiahs*, Princeton: Princeton University Press, 1967.

———. "American Medical Quackery in the Age of the Common Man," *Mississippi Valley Historical Review*, 47:579, 1961.

Yudkin, J., "Sugar and Disease," *Nature* 239:197, 1972.

———. "Sucrose, Insulin and Coronary Heart Disease," *American Heart Journal* 80:844, 1970.

Zavon, M., "MSG: Specific Brain Lesion Damage," *Science* 167:101, 1970.

Index

223

Elizabeth M. Whelan, Sc.D., M.P.H., is a New York City demographer and medical writer. She has taught and researched in the areas of epidemiology, public health and biostatistics. A graduate of Connecticut College, Dr. Whelan has Master of Science and Doctor of Science degrees from the Harvard School of Public Health and a M.P.H. from the Yale School of Medicine, Department of Epidemiology and Public Health. Author of *Sex and Sensibility, Making Sense Out of Sex* and *A Baby? Maybe,* she has contributed articles to *Reader's Digest, Glamour, National Review, Parents' Magazine* and other popular and professional journals.

Fredrick J. Stare, Ph.D., M.D., has been chairman of the Department of Nutrition at the Harvard School of Public Health since its founding in 1942. Dr. Stare is the author of a number of nutrition books including the *Scope Manual on Nutrition* for medical students, now in its third large printing. He is well known for his popular and professional articles in the field of nutrition and to the many readers of his column "Food and Your Health," published by the Los Angeles Times Syndicate.